Conference Board of the Mathematical Sciences

CBMS

Regional Conference Series in Mathematics

Number 104

Collisions, Rings, and Other Newtonian *N*-Body Problems

Donald G. Saari

Published for the
Conference Board of the Mathematical Sciences
by the
American Mathematical Society
Providence, Rhode Island
with support from the
National Science Foundation

NSF-CBMS Regional Research Conference on
The Dynamical Behavior of the Newtonian N-Body Problem
held at Eastern Illinois University, June 9–11, 2002

Partially supported by the National Science Foundation

2000 *Mathematics Subject Classification.* Primary 70F10; Secondary 70F15.

Cover photograph courtesy NASA/JPL-Caltech.

For additional information and updates on this book, visit
www.ams.org/bookpages/cbms-104

Library of Congress Cataloging-in-Publication Data

Saari, D. (Donald)
    Collisions, rings, and other Newtonian N-body problems / Donald G. Saari.
        p. cm. — (CBMS regional conference series in mathematics, ISSN 0160-7642 ; no. 104)
    Includes bibliographical references and index.
    ISBN 0-8218-3250-6 (alk. paper)
        1. Many-body problem—Congresses. 2. Collisions (Astrophysics)—Congresses. I. Title.
II. Regional conference series in mathematics ; no. 104.

QA1.R33 no. 104
[QC174.17.P7]
510 s—dc22
[530.14′4]   2005041205

For two great sons-in-law

Adrian Duffin and Erik Sieberg,

and all of my "N-body" students represented by the first and the obvious,

Neal Hulkower and Zhihong (Jeff) Xia

Preface

This book is the written version of my Conference Board of Mathematical Sciences (CBMS) lectures presented during the week of June 10, 2002, at Eastern Illinois University in Charleston, Illinois. The ten lectures centered on my first and persistent academic love—the Newtonian N-body problem.

While some experts actively participated in the sessions, this conference fully lived up to the intent of the CBMS series in that most of the attendees were graduate students, new-comers to the field, or curious mathematicians wishing to learn something about this fascinating topic. Accordingly, the goals of the lectures quickly changed from a technical presentation appropriate primarily for "experts," to presentations now intended to introduce everyone to the basic structure of N-body systems, to identify certain persistent research themes, and, hopefully, to recruit active participants to this fascinating research area. As such, during each lecture several unsolved research problems were described: some of them are included here.

The new goals for the lectures changed the nature, content, expository tone, and even the subject matter to make the presentations more responsive to the specific interests of the participants while addressing their many questions, For instance, I included more introductory material than originally planned: in retrospect, this was an excellent addition.

The content and approach of this book mimic the changed goals of the lectures; e.g., in addition to new material, you will find discussions intended to develop intuition, introductory material, occasional anecdotes, and descriptions of open problems. To provide cohesion for each chapter, some of the material revolves about unsolved research problems—where the motivating role of the problem may be of more value than the actual problem. In Chap. 1, for instance, much of the discussion is intended to lead to an unresolved issue about the weird dynamics exhibited in the F-ring of Saturn. In Chap. 2, the discussion is tied together via a conjecture involving the diameter of the N-body system. In Chap. 3, the unifying problems involve the important issue of finding certain N-body configurations, which leads to a discussion of the rings of Saturn. In Chap. 4, the issue involves collisions. The concluding Chap. 5 discusses the likelihood of "bad things happening." Everyone, from novices to experts, will find something new.

Some results are new, while others have been presented earlier (e.g., at colloquia, Oberwolfach meetings—particularly several during the 1970s—Midwest Dynamical Systems meetings, a 1983 month long mini-course given in Receife, Brazil while visiting Hildeberto Cabral, in a series of lectures in Paris over 1985-87 hosted by Michael Herman, several informal lectures during 1989 in Barcelona hosted by Jaume Llibre, etc.,) and even advertised as "will appear" in fully intended but never completed papers. In other words, many of these results have not been previously published. As most authors of a book quickly discover, the hard part is not to decide what to include, but what to exclude—particularly if a book is to be eventually completed. (Some of the excluded material probably will appear in [90].)

Other results described in this book come from my earlier papers. The particular journals that published these papers are implicitly acknowledged and thanked via the references. But my expository paper [88] *"A visit to the Newtonian n-body via elementary complex variables"* is extensively used to provide structure and motivation for a couple of the chapters, particularly the introductory one, so I want to explicitly thank the MAA for their permission to use it in this manner.

My deep thanks and appreciation go to Patrick Coulton, the chair for this particular CBMS conference, and my long-time friend Gregory Galprin for inviting me to be the CBMS lecturer and for their efforts to assemble a successful CBMS application. I also thank them for their full and active participation in all lectures and extra sessions that they helped to organize, and for everything they did to make the stay so enjoyable for all of us. I want to thank all of the participants for keeping the workshop sessions so lively! My thanks to the Mathematics Department at Eastern Illinois University for their gracious hospitality. My thanks to Ron Rosier and the CBMS for their program that makes these kinds of lectures possible. Thanks to Neal Hulkower: twice at Northwestern he took my year long course on the Newtonian N-body problem (the first in 1969-70), and he still had both sets of lecture notes! His notes proved to be useful in recovering some of my earlier results and arguments. Also thanks to another student (but I do not recall who it was) who gave me a copy of his notes many years ago.

Irvine, California
January, 2005

Contents

Preface   **v**

1 Introduction   **1**

  1.1 Mars 1

      1.1.1 Motion of Mars 2

      1.1.2 The "far out" planets 4

  1.2 Mercury 6

  1.3 Epicycles 11

  1.4 Chaotic behavior 13

      1.4.1 Newton's method 14

      1.4.2 Period three and circle maps 19

      1.4.3 The forced Van der Pol equations 23

  1.5 The rings of Saturn 26

      1.5.1 Kinky behavior 26

      1.5.2 Λ model 27

2 Central configurations   **31**

  2.1 Equations of motion and integrals 32

  2.2 Central Configurations 34

      2.2.1 Why central configurations are important 36

      2.2.2 Value of λ 40

      2.2.3 Equivalence classes of configurations 42

  2.3 A conjecture and a velocity decomposition 47

      2.3.1 Virial Theorem and the conjecture 47

      2.3.2 The system velocity decomposition 51

      2.3.3 Central configurations and the velocity decomposition 54

      2.3.4 Motion preserving an Euler similarity class 56

2.3.5 Sundman inequality 61
2.4 More conjectures 65
2.4.1 Another conjecture 65
2.4.2 Special cases 67
2.5 Jacobi coordinates help "see" the dynamics 69
2.5.1 Velocity decomposition and a basis 71
2.5.2 Describing ρ'' with the basis 74
2.5.3 "Seeing" the gradient of U 76
2.5.4 An illustrating example 77
2.5.5 Finding central and other configurations 79
2.5.6 Equations of motion for constant I 80
2.5.7 Basis for the coplanar N-body problem 81

3 Finding Central Configurations   **83**
3.1 From the ancient Greeks to 84
3.1.1 Arithmetic and geometric means 84
3.1.2 Connection with central configurations 88
3.2 Constraints 92
3.2.1 Singularity structure of F 94
3.2.2 Some dynamics 97
3.2.3 Stratified structure of the image of F 99
3.3 Geometric approach—the rule of signs 102
3.3.1 The "configurational averaged length" ξ_{CAL} 103
3.3.2 Signs of gradients–coplanar configurations 105
3.3.3 Signs of gradients–three dimensional configurations . . 106
3.3.4 Degenerate configurations 107
3.4 Consequences for central configurations 109
3.4.1 Surprising regularity 109
3.4.2 Estimates on ξ_{CAL} 112
3.4.3 Are there central configurations of these types? 114
3.5 What can, and cannot, be 115
3.5.1 More central configurations 115
3.5.2 Masses and collinear central configurations 119
3.5.3 Masses and coplanar configurations 124
3.6 New kinds of constraints 126
3.7 Rings of Saturn 130
3.7.1 Stability 131
3.7.2 More rings 133
3.7.3 Saturn, and some problems 136

4   Collisions – both real and imaginary   137

  4.1   One body problem 138
        4.1.1   Levi-Civita's approach 140
        4.1.2   Kustaanheimo and Stiefel's approach 141
        4.1.3   Topological obstructions and hairy balls 143
        4.1.4   Sundman's solution of the three-body problem 144
  4.2   Sundman and the three-body problem 147
        4.2.1   Complex singularities? 147
        4.2.2   Avoiding complex singularities 148
        4.2.3   Singularities retaliate 149
  4.3   Generalized Weierstrass-Sundman theorem 150
        4.3.1   A simple case–the central force problem 151
        4.3.2   Larger p-values and "Black Holes" 151
        4.3.3   Lagrange-Jacobi equation 153
        4.3.4   Proof of the Weierstrass-Sundman Theorem 154
        4.3.5   Bounded above means bounded below 159
        4.3.6   Problems 161
        4.3.7   An interesting historical footnote 161
  4.4   Singularities – an overview 162
        4.4.1   Behavior of a singularity 163
        4.4.2   Non-collision singularities 165
        4.4.3   Off to infinity 166
        4.4.4   Problems 170
  4.5   Rate of approach of collisions 172
        4.5.1   General collisions 172
        4.5.2   Tauberian Theorems 173
        4.5.3   Proof of the theorem 179
  4.6   Sharper asymptotic results 184
  4.7   Spin, or no spin? 185
        4.7.1   Using the angular momentum 187
        4.7.2   The collinear case 189
  4.8   Manifolds defined by collisions 191
        4.8.1   Structure of collision sets 192
        4.8.2   Proof of Theorem 4.18 193
  4.9   Proof of the slowly varying assertion 195
        4.9.1   Centers of mass 198
        4.9.2   Back to the proof 201
        4.9.3   The last steps 203

5 **How likely is it?**   **207**
    5.1 Motivation 208
         5.1.1 Idea of proof 209
         5.1.2 Why do we need the Baire category statement? 210
    5.2 Proof: $\mathcal{C}$ is of first Baire category 210
         5.2.1 Finding an appropriate $\mathcal{C}$ subset 211
         5.2.2 A comment about the set of singularities 213
    5.3 Proof: $\mathcal{C}$ is of Lebesgue measure zero 214
         5.3.1 A common collision for $k \geq 3$ particles 214
         5.3.2 Lower dimensions, binary collisions, and other force
                laws 217
    5.4 Likelihood of non-collision singularities 220

Bibliography   **223**

Index   **232**

Chapter 1

Introduction

Simply stated, the "Newtonian N-body problem" is the mathematical study of how heavily bodies move in settings where the dynamics are dictated by Newton's law of motion. In practical terms, this area now includes just about any dynamical system that even remotely resembles Newton's law.

Beyond the insight the subject provides for understanding astronomical issues, the Newtonian N-body problem has historically served as a source of mathematical discovery and new problems. The purpose of this book is to introduce the reader to a selective portion of issues about the Newtonian N-body problem while outlining and describing some open problems.[1]

1.1 Mars

How do the heavenly bodies move? A quick introduction can be provided by using elementary complex variables to describe some simple orbits. The ultimate purpose of this exercise is to show how surprising levels of complexity can arise even in particularly "nice" and "well behaved" settings. Later in this chapter, these orbits are used to describe and motivate an open research problem.

Start with a mystery that most surely bothered generations of school kids: it most certainly troubled me when I was in the fourth grade. It involves the story of Galileo being forced to recant his views that the Sun, rather than Earth, is the center of the solar system. Even a child can appreciate the fact that if the church felt it was necessary to force Galileo to recant, then the stakes in the issue must have been high. But, what

---

[1] A companion book [90] is being prepared that addresses issues other than those described here.

difference does it make if the Sun revolves about the Earth, or the Earth about the Sun? After all, whichever occurs, one forms the center of a circular motion for the other. Why should we care which is which?

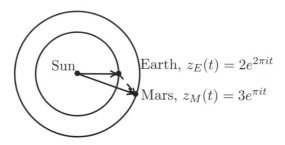

Fig. 1.1. Sun-Earth-Mars coordinates in half-astronomical units

1.1.1 Motion of Mars

To explain the kinds of difficulties that are introduced by an Earth-centered prejudice, start with the Sun as the center of our solar system. A simplified story has Mars approximately 3/2 times (actually, about 1.524 times) as far from the Sun as the Earth, and Mars takes approximately two years (about 687 Earth days) to complete its journey about the Sun.

To keep everything simple, eliminate fractions by replacing the standard astronomical unit (the distance between the Earth and the Sun) with what I call "half-astronomical" units. In the new system, which is depicted in Fig. 1.1, the Earth is two units from the Sun, and Mars is three. Using complex variables, a reasonable description of the motion of the Earth is given by $z_E(t) = 2e^{2\pi it}$ while that of Mars is $z_M = 3e^{\pi it}$.

Finding the orbit of Mars relative to the Earth now is simple; it is

$$z(t) = z_M(t) - z_E(t) = 3e^{\pi it} - 2e^{2\pi it}. \tag{1.1}$$

To describe this orbit, add and subtract the distance to the Sun to obtain

$$\begin{aligned} z(t) \;&= 3e^{\pi it} - 2e^{2\pi it} - 2 + 2 = 2 + e^{\pi it}[3 - 2e^{\pi it} - 2^{-\pi it}] \\ &= 2 + [3 - 4\cos(\pi t)]e^{\pi it}. \end{aligned} \tag{1.2}$$

According to Eq. 1.2, the graph of this equation, as given in Fig. 1.2, depicts the surprisingly complicated orbit of Mars when viewed relative to that of the Earth: it is *a limacon* with a nicely defined loop.[2]

---

[2]In my introductory calculus courses, I often use the trigonometric version of this

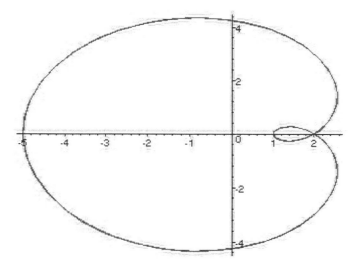

Fig. 1.2. Apparent motion of Mars relative to the Earth

Figure 1.2 makes it clear why the pre-Copernican, Earth-centered preju-dice made it so difficult to predict the motion of the planets and to develop a "Newtonian Theory." For a segment of time on this orbit, everything is regular. Indeed, starting at the point where the loop intersects itself, Mars starts on its long journey moving away from the Earth until eventually it is five half-astronomical units away. (This position corresponds to where Earth and Mars are on opposite sides of the Sun.) The interesting, coun-terintuitive action starts when Mars returns to begin its close approach to the Earth. First, it quickly swoops in a radical plunge toward the Earth. But rather than colliding, Mars suddenly *reverses direction* to swoop out—a motion suggesting that the physics—for some strange reason—suddenly changes to a *law of repulsion* rather than attraction. Finally Mars changes direction once more so that it can repeat its long two-year journey.

Imagine the difficulty in determining the appropriate force law—a law that resembles some form of attraction for most of the journey only to sud-denly become a law of repulsion when Mars approaches Earth, and then reverts back into a law of attraction. Other than resorting to bad jokes about the annoyance of Earthling's politics or their behavior, how does one explain the sudden repulsion of Mars when it starts approaching Earth? In other words, the change of variables from a Earth-centered to a Sun-centered system makes a considerable difference: without it, it is difficult to

---

example to put life into those mandatory reviews of trigonometry. The trigonometric ver-sion just uses double angle formulae; e.g., $(3\cos(\pi t), 3\sin(\pi t)) - 2(\cos(2\pi t), 2\sin(2\pi t)) = (2, 0) + \rho(cos(\pi t), \sin(\pi t))$ where $\rho = 3 - 4\cos(\pi t)$.

even imagine how Newton's laws of attraction could have been developed.

Incidentally, it is easy to observe this retrograde behavior of Mars. Of course, the change in distance between Earth and Mars cannot be detected by the untrained naked eye, but the change in direction—where Mars appears to be moving in one direction, stops and moves backwards, and then stops again to return to its original direction—is quite apparent over the span of several nights. During those periods when Mars approaches Earth to start its dipping behavior, even a casual observer can notice how at a fixed time each night the position of Mars swings to define, over a period of days, a compressed "Z."

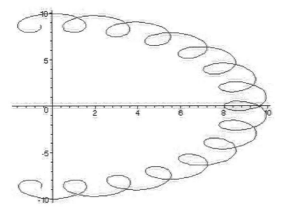

Fig. 1.3. Apparent orbit of a planet 9 times farther from the Sun

While the apparent motion of Mars offers surprising behavior, the orbits of the planets farther from the Sun adopt a much more complicated appearance with the several loops as indicated in Fig. 1.3. This figure depicts the apparent behavior of a planet nine AU away from the Sun: a distance that is a bit short of Saturn's actual orbit. Rather than developing a complicated version of the above description, a different elementary approach is described next.

1.1.2   The "far out" planets

Consider the circular orbit of a far-out planet—Mars, Saturn, or beyond—given by $z_P(t) = ae^{\alpha \pi i t}$ where the value of $a \geq 3$ defines the distance from the Sun in our half-astronomical units: the α values are discussed below. After expressing this

$$z(t) = z_P(t) - z_E(t) = ae^{\alpha \pi i t} - 2e^{2\pi i t}, \qquad (1.3)$$

orbit of the planet relative to the Earth in the usual complex variable form of $z(t) = r(t)e^{i\theta(t)}$, a way to determine whether the orbit is moving in a clockwise or counter-clockwise manner (relative to the Earth) is to examine the sign of $\theta'(t)$.

The sign of $\theta'(t)$ is the imaginary part of $(\ln z_P(t))' = \frac{z_P'}{z_P} = \frac{r'}{r} + i\theta'$. But since

$$(\ln z_P(t))' = \frac{z_P'}{z_P} = \frac{\pi i(a\alpha - 4e^{(2-\alpha)\pi it})}{a - 2e^{(2-\alpha)\pi it}}, \tag{1.4}$$

it follows from the form of the numerator that the sign of θ' must change periodically whenever $a\alpha < 4$.

The reason this $a\alpha < 4$ inequality must hold for all of the planets that are farther from the Sun than the Earth is *Kepler's third law*. This law asserts that

$$a^3\alpha^2 = k \tag{1.5}$$

where k is a constant. Consequently, $a\alpha = (\frac{k}{a})^{1/2}$ is a decreasing function of a: remember, a is the distance of the planet to the Sun. Thus, for a planet sufficiently far from the Sun, we must expect its orbit to experience loops when expressed relative to the Earth. According to Eq. 1.4, the loop occurs whenever the distance between the Earth and the planet decreases toward a (local) minimal value. But because those far-out planets take from decades to a couple of Earth centuries to circle the Sun,[3] it follows that their apparent orbits must exhibit many loops.

A natural related question, which is needed for later purposes, is to determine how far a planet must be beyond the Earth so that its apparent orbit has a loop. Using the units of the Earth, $a = 2, \alpha = 2$, we have that $k = 32$ for Eq. 1.5. Thus, $a^3\alpha^2 = 32$, or the crucial parameter has the value $a\alpha = [32/a]^{\frac{1}{2}}$. Because apparent loops occur when $a\alpha < 4$, it follows that these loops occur when $[32/a]^{\frac{1}{2}} < 4$, or when $a > 2$. Restated in words,

> *the apparent motion of any planet that is farther from the Sun than the Earth has a loop.*

Of course, this assertion holds for all bodies governed by Newton's equation: this fact plays a key role in the discussion about the rings of Saturn given in the last section of this chapter.

Notice how this simple argument just describes a circular uniform motion relative to another circular uniform orbit. The importance of this comment

---

[3]While Venus takes only about 224 Earth days to circle the Sun, Jupiter takes 4332 (about 11.9 Earth years), Saturn 10,760 (about 29.5 years), Uranus 30,685 (about 84 years), Neptune 60,190 (about 165 years), and Pluto 90,800 days (about 249 years).

derives from the reality that all sorts of circular motions arise in the N-body problem. Consequently, it is reasonable to anticipate that loops and complicated orbits of this type are commonplace. To illustrate with another example, I now turn to the orbit of Mercury.

1.2   Mercury

Mercury, our smallest planet, is only slightly larger than our Moon. Even though this planet was known to the ancient Egyptians, its proximity to the Sun has hampered all attempts to explore Mercury with either telescopic observations or space missions.[4] Yet, enough is known about Mercury to allow it to play an important role in the development of celestial mechanics. We know, for instance, that the perihelion of Mercury (i.e., the closest approach of Mercury to the Sun) deviates 43" of arc per century from that predicted by Newton's laws. What a stunning assertion! When you consider only 43" of arc *per century* you have to join me in being impressed by the precision attained by our nineteenth century colleagues!

This deviant behavior has encouraged all sorts of searches including speculation about the possible existence of another planet called "Vulcan."[5] What a delightful notion: could it be that someone would discover (again!) a new planet strictly through *mathematical computations*? Einstein spoiled the fun by showing that this effect could be explained strictly in terms of his theory of relativity.

Moving on to the orbit of Mercury, we now know that this planet takes

---

[4]Mercury was visited by NASA's Mariner 10 in March of 1974 where the pictures sent back from its three approaches reveal a planet with plains of frozen lava and a surface pockmarked with craters: the planet resembles our Moon. The Mariner photos also discovered the Caloris Basin; a basin that suggests one of the most cataclysmic events in our planetary system. Mercury is not an inviting destination.

[5]It is highly doubtful that Vulcan exists, yet its colorful history involved important individuals. For instance, the French mathematician Urbain Le Verrier ensured a place in history by using computations about how the path of Uranus deviated from Newtonian predictions to predict the existence and position of Neptune; he did this before Neptune was even observed. Using similar reasoning, in 1860 he wondered whether the deviation of Mercury's perihelion indicated the existence of another planet or an asteroid belt. Accelerating the chase for the discovery of the new planet, which Le Verrier christened "Vulcan," were the claims of the amateur astronomer Lescarbault that he sighted a spot— a planet?—near the Sun. But Le Verrier showed that even if this sighting were the speculated Vulcan, its orbit would not explain Mercury's perihelion problems. Again in 1878 two reputable astronomers, Watson and Swift, suspected they saw "stars" that might be the elusive Vulcan. With the exception of faint objects observed near the Sun during a 1970 solar eclipse, if Vulcan really exists, it has successfully remained hidden.

about 90 Earth days (actually, 87.97) to circle the Sun. If you check the books in astronomy written prior to 1965, they reflect the earlier belief that Mercury took about the same length of time—about 90 Earth days—to rotate on its axis. If this were true, then, as these books asserted, the same face of Mercury would always face the Sun. This would constitute, of course, a phenomenon similar to where the same face of the Moon always faces the Earth.

In 1965 radio astronomers visited this long standing belief. Using the Arecibo radio telescope based in Puerto Rico, they discovered that the sideral rotation (the length of time it takes a planet to revolve once as measured against a fixed background) was about 60 days (actually 59, or two thirds of the length of a Mercury year). In turn, this change in the rotational period significantly shortens a Mercury solar day from eternity to about 176 Earth days, or exactly two Mercury years.

Our interest in these figures is that they identify another astronomical setting where it is reasonable to consider the orbit of a rotating object relative to a point on another rotating object. Stated in simple terms, if someone lived on Mercury, what would the orbit of the Sun look like?

To answer to this question, we need to use a more accurate description for the orbit of Mercury. A sharper approximation for the orbit of any planet is obtained by treating it as an ellipse with eccentricity ϵ rather than a circle where $\epsilon = 0$. For instance, the Earth's orbit is fairly circular as reflected by its eccentricity of $\epsilon = 0.0167$, while the more elliptical orbit of Mercury is manifested by the twelve-fold larger $\epsilon = 0.2056$ value. Indeed, Mercury's more extreme elliptical nature is captured by the difference between its perihelion distance of 0.308 AU and its aphelion (its largest distance from the Sun) of 0.466 AU.

The position of any planet on its ellipse is given by

$$r_P(\theta) = \frac{a}{1 - \epsilon \cos(\theta)} \approx a(1 + \epsilon \cos(\theta)) \tag{1.6}$$

where a is a positive constant (the length of the semi-major axis of the ellipse) and $\theta(t)$ is the angular position of the planet relative to a reference line.[6] This equation has the complex variable representation

$$z_P(\theta) \approx a(1 + \epsilon \cos(\theta))e^{i\theta}. \tag{1.7}$$

By measuring time in Earth years and using the fact Mercury takes about 60 days, or one-sixth of an Earth year, to rotate on its axis, the

---

[6]For a quick introduction into the basics of the two-body problem, I recommend the first chapter of Pollard's book *Celestial Mechanics* [60].

rotating motion of Mercury about its axis can be approximated by $e^{12\pi it}$. Thus the apparent position of the Sun (relative to a position on the surface of Mercury) can be represented by the product

$$Z_S(t) = -z_P(\theta(t))e^{-12\pi it}. \tag{1.8}$$

In turn, the angular position of the Sun, given by the argument of $Z_S(t)$, is

$$\text{Arg}(Z_S(t)) = \text{Arg}(-z_P(\theta(t))e^{-12\pi it}) = \theta(t) - 12\pi t. \tag{1.9}$$

According to this expression, the Sun's apparent motion changes direction whenever $(\text{Arg}(Z_S(t)))'$ changes sign; that is, whenever θ' passes through the value of 12π. The reason the Sun's orbit relative to a position on Mercury must experience this directional change comes from *Kepler's second law* that captures the angular momentum: this law asserts that

$$r_P^2\theta' = C \tag{1.10}$$

where C is a constant. What we learn from Eq. 1.6 is that if ϵ is sufficiently large, as it is for Mercury, then it is unreasonable to approximate $r_P(t)$ by a constant. In turn, according to Kepler's second law (Eq. 1.10), $\theta'(t)$ cannot be approximated by a constant, so $\theta(t)$ cannot be approximated by uniform motion. In particular, when $r_P(t)$ is at perihelion, θ' achieves its maximum value; when $r_P(t)$ is at apihelion, θ' attains its minimum. This assertion matches intuition gained from those high school physics exercises of swinging a weight on a string where pulling in the string (i.e., r_P is made shorter) makes the object move faster.

Using Kepler's second law and Eq. 1.6, we have that

$$\theta'(t) = \frac{n}{(1 - \epsilon\cos(\theta))^2}$$

where the new constant n—a combination of a and C—is called the "mean motion." For Mercury, $n = 8\pi$, so, by setting

$$\theta' = \frac{8\pi}{(1 - \epsilon\cos(\theta))^2} = 12\pi, \tag{1.11}$$

it follows immediately that the apparent motion of the Sun moves in different directions depending on whether $(1 - \epsilon\cos(\theta))^2$ is greater than, or less than, $\frac{2}{3}$. The first condition requires $r_P(\theta)$ to have a sufficiently large value, the second requires $r_P(\theta)$ to have a sufficiently small value. In turn, this means that we should expect a reversal in direction to occur whenever Mercury moves closer to the Sun.

What remains is to seen whether the orbit of Mercury ever experiences a reversal; namely, can $\theta' - 12\pi$ change sign? The answer follows by examining what happens for $\theta = 0$ —at perihelion, and $\theta = \pi$—at aphelion. A direct computation shows for Mercury that the θ' values at perihelion and aphelion are, respectively,

$$\theta' = \frac{8\pi}{(0.7944)^2} = 12.677\pi > 12\pi, \quad \frac{8\pi}{(1.2056)^2} < 12\pi.$$

Consequently, the apparent orbit of the Sun relative to Mercury must experience reversals.

More specifically, whenever Mercury moves toward its closest approach to the Sun, the Sun's apparent motion *reverses direction*. But there are two Mercury years for each Mercury day,[7] so this unexpected phenomena happens "twice a Mercury day." To add some drama to the description, as indicated in Fig. 1.4, at some location the Sun will rise in the east, only to almost immediately *set again in the east* for a period of time. Then, the Sun does rise a second time in the east for the long Mercury day until it finally appears to set in the west. The correct word is "appears" because shortly after the Sun sets, it *rises again in the west* for a short time, only to finally settle in the west for the long Mercury night.[8]

---

[7]Mercury's day-year ratio has intrigued mathematicians. Tom Kyner [32] probably was the first to explore this resonance effect by using dynamical systems to show how the orbit is "trapped." Kyner introduced his results at a 1969 conference on mathematical astronomy held in Sao Paulo, Brazil. Amusingly, Kyner presented his paper the day after the first and (probably) only time he observed Mercury: this was at a social gathering held at the Sao Paulo Observatory. (Most of us conference participants were mathematicians, so visiting an observatory was a novelty.) Later C. Robinson and J. Murdock [70] extended the mathematics. But, if this Mercury rotation problem is interesting, a greater challenge comes from the planet Venus. Years ago friends at JPL told me that prior to radio astronomy, astronomers knew nothing about the rotation of Venus. What they discovered through modern technology was surprising: Venus moves in a *retrograde* motion. Why this "backwards" rotation? I expect that an explanation will involve some fascinating mathematics. In any case, it provides an exercise that I leave to the reader: what is the apparent behavior of the Sun for someone on Venus?

[8]I was delighted when in 1991, Chris Fang-Yen, a high-school student supervised by Stan Wagon in a summer research program at the (now defunct) Geometry Center at the University of Minnesota, sent me his project entitled "Sunrise and Sunset on Mercury." After Wagon read my paper (Saari [88]) where I described this phenomenon, he asked Fang-Yen to simulate the motion. In doing so, Fang-Yen discovered that there does not exist a *fixed position* on Mercury where this double-dip behavior can be observed. The reason is that the Sun is too small and, as distinctly suggested by Fig. 1.4, the dips in the apparent orbit are too tight relative to the radius of Mercury for a Mercurian to see over the horizon. Consequently, to observe this phenomenon, our Mercurian would have

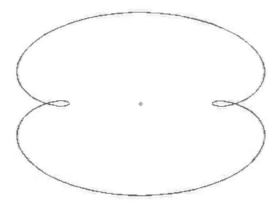

Fig. 1.4. Apparent behavior of the Sun from Mercury

It is clear from Eq. 1.11 that this strange phenomenon is due to the large eccentricity of Mercury. A direct computation, which just involves finding the ϵ value where

$$\theta' = \frac{8\pi}{(1-\epsilon)^2} \leq 12\pi,$$

proves that the apparent motion of the Sun would have no direction reversal had the orbit of Mercury been circular enough so that $\epsilon \leq 0.1835$.

What we have observed in this and in the last section is how different but seemingly unremarkable orbits of two bodies about a central body, or even the orbit of one body about another but with a sufficiently large eccentricity, converts the actual motion into an apparent one with surprisingly complex behavior. A more important observation is that this effect is a direct consequence of describing one circular motion relative to another circular motion. Consequently, we must anticipate this behavior to be reasonably common in celestial mechanics. The interesting message, which helps identify new research issues and opportunities, is that the closer the bodies approach one-another in a Newtonian system, not only does the force between them increase, but it is possible to have an apparent reversal of direction. Thus, as indicated in Figs. 1.2, 1.3, and 1.4, expect surprising and recurring changes in the gravitational forces to arise even in seemingly well-behaved settings. This observation plays a central role in Sect. 1.5.

---

to move to another location. But this should be no problem because he would have "all Mercury day" to do so.

1.3 Epicycles

With an Earth centered system, how does one recognize and represent the motion of the planets? This problem was crucial for astronomers of ancient time. After all, researchers of that epoch needed accurate representations in order to construct astrology tables. Don't scoff at these efforts because, quite frankly, the money paid for these commissioned tables could be viewed as the NSF research funding of that time.

The accuracy needed for this representation problem was achieved through the predictive planetary theory developed by Ptolemy. To appreciate the genius of his work, recall that any theory must adjust to the prejudices of the day; the restrictions facing Ptolemy were the monumental ones established by Aristotle. As a quick, maybe overly simplistic review, Aristotle believed

(1) that the Earth was the center of the universe, and

(2) that the circle and uniform circular motion were the most virtuous figure and motion.

Of course, since virtue is located in the "heavens,"

(3) any description of the motion of the planets must be described in terms of uniform circular motion about the Earth.

Fine, but how?

In his *Almagest,* written around 130 A.D., Ptolemy resolved the problem of describing the position of the planets with circular, uniformly moving motion by putting forth his ingenious epicycle approach. The clever idea is that the point indicated by a designated point on the motion of the first circle, the *deferent,* does not represent the location of the planet. Instead, the point merely locates the center of a *second* circle that also is spinning with uniform motion. The location of the planet, then, is given by the moving point on the second circle—the epicycle. (See Fig. 1.5.)

Today this approach may seem to be hopelessly naive. But remember that variations of Ptolemy's theory dominated astronomy for more than a millennium—this is an incredibly long period of time for any scientific theory. Even Newton's theory did not enjoy such a long reign before being challenged by Einstein's relativity, and who knows how long it will take until a serious challenge will force Einstein's theory to be replaced.

The long success of Ptolemy's approach can be understood, again, in terms of elementary complex variables. Let a_j be the radius of the jth circle where the uniform motion takes c_j Earth years, $j = 1, 2$, to complete one revolution. This means that the motion of a planet as described by epicycles is given by

$$z_P(t) = a_1 e^{b_1 \pi i t} + a_2 e^{b_2 \pi i t} \text{ where } b_j = \frac{2}{c_j}, \ j = 1, 2. \qquad (1.12)$$

This expression should be familiar: by comparing Eq. 1.12 with the Earth centered expressions Eqs. 1.1, 1.3, we discover that the epicycles can be treated as roughly capturing the conversion of a Sun centered representation of the location of the planets into an Earth centered one. Namely, let the deferent be the location of the Sun relative to the Earth, and let the epicycle describe the location of Mars relative to the Sun. (The actual descriptions were more complicated.) No wonder Ptolemy's approach proved to be so successful!

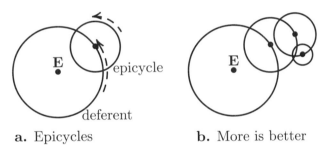

a. Epicycles   **b.** More is better

Fig. 1.5. Epicycle structure

Of course, the demanding needs of astronomers and astrologers to obtain even sharper representations and to correct for accumulated error required a sharper, extended theory. One approach is obvious. Instead of treating a point on the epicycle as the location of a planet, interpret it as the center of still another rotating circle. As indicated in Fig. 1.5b, even that point could be treated as the center for still another rotating circle. Imagine: these attempts to find still better theoretical representations could be the source of an infinite number of Ph.D. theses. By adding more and more terms, we obtain the expression

$$z_P(t) = \sum_j a_j e^{b_j \pi i t}. \qquad (1.13)$$

Again, we all accept that the epicycle approach is hopelessly naive. After all, N-body motion is much more complicated where modern theories are accompanied with terms such as "quasi-periodic" or "almost periodic" motion. What is quasi-periodic motion? It is represented by a finite sum in

Eq. 1.13 while almost periodic motion is given by a converging infinite sum. In other words, once we recognize that epicycles geometrically represent motion developed in modern theory, we gain added respect for Ptolemy's insight.

As a brief aside, Eq. 1.13 describes a Fourier series when the b_j's are integer multiples of a specified number. But quasi-periodic and almost periodic motion imposes no constraints on the b_j values. This amazing theory of almost periodic motion was essentially developed by one person, Harald Bohr [6, 7], the younger brother of Niels Bohr.

Before describing more structure of these motions, first recall the story of a mathematician asked to give a general audience talk about mathematics to the parents and teachers of his daughter's elementary school starting off with, "Let X be a non-separable Hilbert Space." The delight of this self-deprecating joke—the kind many mathematicians enjoy—is that many of us cannot quickly provide an example of such a space. To do so while illustrating some of the mathematical structure Bohr developed, define the inner product for complex valued functions on the real line by

$$(f, g) = \lim_{T \to \infty} \frac{1}{2T} \int_{-T}^{T} f(t)\overline{g(t)} \, dt,$$

and define a Hilbert space in the normal manner. It is easy to show that an orthonormal basis for this space is $\{e^{i\lambda t}\}$ for $\lambda \in R$. As this space—the natural home for epicycles, quasi- and almost periodic motion—has an uncountable basis, it is a natural example of a non-separable Hilbert Space.

1.4 Chaotic behavior

To relate the above story about the behavior of Mars, Mercury, and the other planets to an open research question, allow me a slight digression to describe, in what is intended to be a reader friendly introduction to certain basic concepts from chaotic dynamics. Since this book describes Newtonian mechanics, it is appropriate to describe this behavior in terms of Newton's method for finding zeros of a function.[9] After introducing certain basic points, I briefly touch on that well-known "Period Three implies Chaos" paper by T-Y Li and James Yorke [33].

---

[9]The material and exposition for this section comes from Saari and Urenko [93] and from Saari [89].

Let me stress that I have no intensions to fully analyze Newton's method nor to describe the subtle features of chaotic dynamics. The intent is strictly to help develop intuition by suggesting what features are indicators of the potential complexity of motion for the Newton's N-body problem. Hopefully this brief, intuitive description will entice readers not familiar with this topic to learn more about these standard tools for the study of the N-body problem and celestial mechanics. More complete descriptions are readily available in books such as Alligood, Sauer, and Yorke [2], Devaney [14] and Robinson [68, 69].

1.4.1   Newton's method

To review, Newton's method for finding a zero of a function $y = f(x)$ starts with an initial guess, x_1, and finds an improved estimate. As indicated in Fig. 1.6, this next choice is found by replacing the specified function with its straight line approximation passing through $(x_1, f(x_1))$. The zero of this linear equation $y = f(x_1) + f'(x_1)(x - x_1)$, denoted by x_2, is the new estimate. If x_2 is not a zero, the process continues. The iteration process has the expression

$$x_{n+1} = x_n - \frac{f(x_n)}{f'(x_n)}. \tag{1.14}$$

But rather than using the Eq. 1.14 analytic expression, the geometry of this process as indicated in Fig. 1.6 suffices for our purposes.

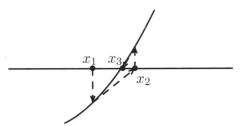

Fig. 1.6. Newton's Method

As it is easy to establish, and as indicated in Fig. 1.6 where the third iterate x_3 already has nearly located the zero, once an iterate is sufficiently near a zero, the process converges to this zero. But as this convergence assertion describes only highly local behavior, it is natural to question what happens globally. Figure 1.7, for instance, represents a polynomial of degree five. The goal is to determine the general dynamical behavior of Newton's method when applied to this function.

Stated in words, other than converging to a zero, what can go wrong?

What else can happen with Newton's method? A little experimentation suggests that there exist period two points: points where Newton's method bounces between them forever. Other points have more serious consequences because they cause Newton's method to cease to exist. In Fig. 1.7, four of these points, depicted by bullets, identify the function's critical points. Since $f'(x_j) = 0$ at the critical points, the horizontal linear approximation never meets the x-axis, so the next iterate is not defined.

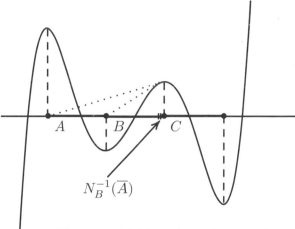

Fig. 1.7. Fifth order polynomial

These four critical points divide the line into five intervals; the two extreme regions are unbounded and the three bounded intervals are labelled A, B, and C. It follows from the properties of polynomials (i.e., the convexity of the curve) that if an iterate ever lands in one of the two unbounded regions, the subsequent iterates converge to the zero of that region. Consequently, all unusual and nonconvergent properties of Newton's method must be limited to the $A \cup B \cup C$ region.

Finding nonconvergent properties

The natural and traditional way to discover "nonconvergent" properties of Newton's method is to examine the behavior of the orbit $x_1, x_2, \ldots, x_n, \ldots$ and experiment with the choice of x_1. If, for instance, iterate x_{101} ends up near x_1, then it is reasonable to expect from continuity considerations that a slight and suitable change in the choice of x_1 will force $x_1 = x_{101}$. Stated in other terms, this means that we should expect the existence of a 100-period orbit. This is an overly simplified, but not inaccurate, description of how various properties about Newton's method were obtained. Have no doubts: while this approach necessarily uncovers only limited conclusions, it can be technically very difficult.

To motivate an alternative way to examine Newton's method, first remember that anything new or interesting requires keeping all iterates within
the region $A \cup B \cup C$. To introduce the approach, after determining an orbit
$\{x_1, x_2, x_3, \ldots, x_n, \ldots\}$, replace each iterate with a label identifying the interval in which it belongs. For instance, if $x_1 \in C, x_2 \in A, x_3 \in B, \ldots$ then
the initial condition x_1 and its iterates defines the sequence

$$g_f(x_1) = \{C, A, B, \ldots\}.$$

This listing of intervals is not random—it is specified by the dynamic process—
so give it status by calling it a *word* or an *itinerary*.

Let $U^3 = \{A, B, C\}^N$ be the *universal* set where N is the set of natural
numbers. In words, U^3 consists of the uncountable number of all possible
sequences that can be constructed with the entries A, B, and C: call it the
universal set. With this notation, a word generated by the initial condition
x_1 and function f is given by a mapping[10]

$$g_f : A \cup B \cup C \to U^3. \tag{1.15}$$

A way to measure the complexity of Newton's method applied to function
f is to determine all entries—all words—in the *dictionary*

$$\mathcal{D}_f = \{g_f(x) \in U^3 \,|\, x \in A \cup B \cup C\}.$$

After all, common sense dictates that if $\mathcal{D}_f$ is a large subset of U^3, then
Newton's method admits rich, complex, chaotic dynamics. But if $\mathcal{D}_f$ has
only a limited number of words, then it corresponds to a relatively benign
dynamic. These comments suggest that a way to measure the complexity
of the admissible dynamics is to determine the dictionary $\mathcal{D}_f$. Notice the
changed emphasis. Rather than finding particular properties of Newton's
method, the ambitious new goal is to completely characterize and catalogue
all long term dynamical properties—it is to identify all itineraries in $\mathcal{D}_f$.

The surprising complexity of Newton's method

As shown next (from Saari and Urenko [93]), Newton's method can be highly
complex in terms of this complexity measure.

---

[10]Some choices of x will have their iterates in an unbounded region. Either ignore them
as I am doing, or handle them in an obvious manner.

Theorem 1.1 *(Saari and Urenko [93]) For a fifth-order polynomial $y = f(x)$ with distinct real roots*

$$\mathcal{D}_f = U^3.$$

This theorem means that we can choose any sequence consisting of the letters A, B, and C—even a sequence generated by rolling a die—and we are assured that there exists an initial iterate in the specified first interval where the jth Newton method iterate lies in the region specified by the jth entry of the sequence; $j = 1, 2, \ldots$. As indicated below, this means that periodic orbits of any length must exist, as well as far more complex behavior.

The theorem asserts that g_f, defined in Eq. 1.15, is surjective. The method to establish the surjectivity of g_f uses an "iterated inverse image" approach that I illustrate with the sequence

$$w = \{B, A, C, C, A, \ldots\}. \tag{1.16}$$

The approach is to keep refining the set of initial iterates that accomplish each portion of the proposed itinerary.

To see how to do this, let N_k be the portion of Newton's map restricted to interval k where $k = A, B, C$. By using the inverse mappings N_k^{-1}, we have, for instance, that the set of initial iterates starting in the specified first interval B of the Eq. 1.16 sequence and ending in the closed second specified interval $\overline{A}$ is given by $N_B^{-1}(\overline{A})$. The key fact is that N_k maps interval k onto $(-\infty, \infty)$. To see why this is true, notice that as x moves closer to the left endpoint of the bounded interval k, $N_k(x) \to \infty$, while as x moves toward the right-hand endpoint, $N_k(x) \to -\infty$. The conclusion now follows from the continuity of N_k on interval $k = A, B, C$.

Because $N_k : k \to (-\infty, \infty)$ is surjective for $k = A, B, C$, it follows that $N_B^{-1}(\overline{A})$ is a closed subset of B. Actually, as indicated in Fig. 1.7 with the dotted lines, this set is easy to roughly determine: just find the inverse Newton image of the two endpoints of interval A. That is, in interval B find tangent lines to the graph of $y = f(x)$ that terminate on the endpoints of A. The corresponding x values define the endpoints of $N_B^{-1}(\overline{A})$.

Set $N_B^{-1}(\overline{A})$, which identifies all points starting in B that are mapped to A, is much more than we want. After all, our interest is to land only on those points in A that are then mapped to C: we are only interested in the points that are mapped to $N_A^{-1}(\overline{C})$. This set $N_A^{-1}(\overline{C})$ is determined in precisely the same fashion. Thus, refining our set of initial conditions to the closed set $N_B^{-1}(N_A^{-1}(\overline{C}))$ identifies all initial iterates in B that are mapped to A and then to C.

The approach now is obvious. To find all initial points satisfying the future specified in the Eq. 1.16 sequence, continue this iterated inverse image approach to obtain the nested sequence of bounded, closed subsets

$$\overline{B} \supset N_B^{-1}(\overline{A}) \supset N_B^{-1}(N_A^{-1}(\overline{C})) \supset \cdots \supset N_B^{-1}(N_A^{-1}(\ldots N_k^{-1}(\ldots)\ldots)) \supset \ldots.$$
(1.17)

By construction, a point in the intersection of all subsets in Eq. 1.17 must satisfy the specified future. But, by appealing to standard results from a first course in real analysis about a nested sequence of bounded, closed sets, we know that such a point must exist. Thus, whatever the envisioned future, it can occur.

Sensitivity and Cantor sets

This construction provides an intuitive description of the source of several of the phrases—"sensitivity to initial conditions," "Cantor sets," etc.— common to this area. To start with the sensitivity phrase, notice how the *expanding* nature of N_k ensures that when N_k is examined in the inverse direction, the inverse image $N_B^{-1}(\overline{A})$ must be a small subset of B. Indeed, by checking Fig. 1.7, it is clear that $N_B^{-1}(\overline{A})$ is a very tiny subinterval.

While set $N_B^{-1}(\overline{A})$ is quite small, its subset $N_B^{-1}(N_A^{-1}(\overline{C}))$ is much smaller. But by construction, this $N_B^{-1}(N_A^{-1}(\overline{C}))$ subset contains *all points* starting in B that Newton's method moves to A and then to C. After the points arrive in C, what happens next? Anything you want: this comment is a direct consequence of the image of N_C being $(-\infty, \infty)$. Stated in words, this surjectivity of N_C along with the small size of $N_B^{-1}(N_A^{-1}(\overline{C}))$ means that even the slightest difference between points in $N_B^{-1}(N_A^{-1}(\overline{C}))$ could result in radically different futures: the dynamic behavior is "sensitive with respect to initial conditions."

Similarly, for each extension of $\{B, A, C, \ldots\}$, each step of the iterated inverse image approach identifies all points that eventually are mapped onto the next specified interval. Included among these points are open intervals that converge to the zero in this interval or in one of the two unbounded regions. Thus, to construct the set of points of nonconvergence of Newton's method, open sets need to be excised at each step—just as in the construction of the "middle thirds" Cantor set. In other words, expect Cantor sets.

With the exception of the behavior of Newton's method on the unbounded regions, nothing in this description restricts the story to polynomials with five real, distinct roots, or even to polynomials. This means, for instance, that the same phenomenon will arise in any polynomial with at least four real and distinct roots. Just imagine what happens with Newton's method applied to $y = \cos(x)$ with the infinite number of symbols!

All that is needed for this story is that the map incurs an *expansion*—given by the fact that the image of each continuous N_k includes $A \cup B \cup C$—and a *recurrence* effect—captured by those $A \cup B \cup C$ points that are mapped back to $A \cup B \cup C$. It is this *expansion* and *recurrence* combination—a common combination for celestial mechanics— that provides the interest for the N-body problem. Namely, anticipate complex, chaotic behavior in the N-body problem.

1.4.2 Period three and circle maps

To reinforce the basic notions, they are described again using the "Period three implies chaos" title of the influential 1975 paper written by Tien-Yien Li and James Yorke [33]. Beyond the nice mathematics, this paper has historical interest because it is where the term "chaos" originated. As the story goes, after the paper was accepted, the editor of the journal asked the authors to change the title to something mathematically more acceptable and descriptive—maybe something such as "Period three implies topological transitivity." How dull. Fortunately Yorke remained firm in his intent to retain the original title, and the term "chaos" was coined.

Sarkovskii sequence

Unknown to Li and Yorke, eleven years earlier A. N. Sarkovskii [99] published a remarkable and stronger result that if a continuous mapping from the line to the line had a period three orbit, then it also has periodic points of any period. More precisely, Sarkovskii proved for the following sequence, now called the *Sarkovskii sequence*,

$$
\begin{array}{cccccc}
3, & 5, & 7, & 9, & 11, & 13, & \ldots \\
2 \cdot 3, & 2 \cdot 5, & 2 \cdot 7, & 2 \cdot 9, & 2 \cdot 11, & 2 \cdot 13, & \ldots \\
2^2 \cdot 3, & 2^2 \cdot 5, & 2^2 \cdot 7, & 2^2 \cdot 9, & 2^2 \cdot 11, & 2^2 \cdot 13, & \ldots \\
\cdots \\
\cdots \\
2^n \cdot 3, & 2^n \cdot 5, & 2^n \cdot 7, & 2^n \cdot 9, & 2^n \cdot 11, & 2^n \cdot 13, & \ldots \\
\cdots \\
\cdots \\
\cdots & 2^n & \cdots & 2^3 & 2^2 & 2^1 & 1
\end{array}
$$

$$(1.18)$$

that a continuous mapping from the line to itself with a periodic point of period k also has periodic points for each period that follow k in the above listing. Thus, for instance, if such a mapping has a period $k = 3$ point, then, because all positive integers follow 3 in this Sarkovskii sequence, initial points can be found where the same map has a period 1,000,345 point, or a fixed point, or points of any other possible period.

Period three maps

Suppose f is a continuous map from the line back to the line that has the period three point

$$f(x_1) = x_2, \quad f(x_2) = x_3, \quad f(x_3) = x_1.$$

Choose some ordering of these three points, it does not matter what it is, and then plot the three $(x_j, f(x_j))$ points. The assertion is that any way these points can be connected to form a graph of a continuous function, the resulting function admits periodic points of any period along with far more complex behavior. The choice of the mapping selected for Fig. 1.8 provides an unimaginative but minimal straight-line way to connect these points where the selected ordering is $x_1 < x_2 < x_3$.

By definition, this Fig. 1.8 mapping takes x_1 and x_2—the endpoints of the interval A—respectively to points x_2 and x_3—the endpoints of interval B. It now follows from continuity that this mapping—however it may be drawn—must, at the minimum, map interval A onto interval B.

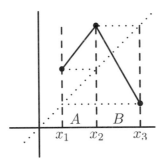

Fig. 1.8. A period three map

Similarly, since the endpoints of B, given by x_2 and x_3, are mapped respectively to x_3 and x_1—these are the endpoints of $A \cup B$—the period-three mapping must experience an *expansion* effect where interval B is reversed and then stretched to be mapped to $A \cup B$. Cataloguing this information as

$$f(\overline{A}) \supset \overline{B}, \quad f(\overline{B}) \supset \overline{A \cup B},$$

it now follows by using the above "iterated inverse image" argument that this f admits a rich variety of different dynamics. In particular, at a minimum ("minimum" because by drawing more expansive maps rather than just connecting the three points, wilder possibilities become possible), we know that any sequence is a word if it satisfies the conditions where

> an A always is followed by a B; a B can be followed by an A or a B.

For any such sequence, we know there exists an initial iterate satisfying this future. Again, it is the combination of expansion and recurrence that creates the complicated dynamics.

Presumably, a period-five point would be given by a sequence that continually repeats the block $ABBBB$, or maybe the block $ABABB$, while a period seven point is obtained from repeated blocks of, say, $ABBABBB$. More complicated orbits that avoid having any periodicity, but skirt arbitrarily close to various period points, are represented by sequences where no block ever repeats itself. One example of this is

$$ABABBABBBABBBBABBBBBA\ldots$$

where each A is followed by even longer sequence of B's.

Circle maps

The preceding paragraph cautiously states that "Presumably, a period-five point would be given by a [repeating] sequence." It does, and a simple way to prove this assertion can be illustrated by using a mapping from the circle to the circle.

Actually, we already have analyzed a mapping $f : S^1 \to S^1$: a mapping from the circle to the circle. This is Newton's method because, by the usual trick of adding a point at infinity, the infinite line becomes a circle where the infinity point is the North Pole. With this representation, Newton's method is continuous as the critical points of a function are mapped to the North Pole.

Suppose we have a simple continuous mapping $f : S^1 \to S^1$ that wraps around the circle twice: for simplicity, consider $f = e^{2\theta}$, $0 \le \theta \le 2\pi$. The goal is to demonstrate that this mapping has periodic points of all periods. By slicing the circle open at the North Pole and flattening it into a line interval $[0, 2\pi)$, the graph becomes as displayed in Fig. 1.9a.

Subintervals A and B are as designated in Fig. 1.9a, and the figure shows that

$$f(\overline{A}) = \overline{A \cup B}, \quad f(\overline{B}) = \overline{A \cup B}.$$

According to the above argument, it now follows that any sequence consisting of A's and B's can be realized by some initial iterate. Presumably, this means that a sequence repeating the block AB is a period-two point while the one repeating, say, the block ABB, is a period three point. A simple argument using the graph shows that, indeed, these are periodic points.

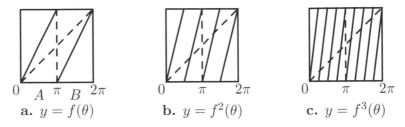

a. $y = f(\theta)$ **b.** $y = f^2(\theta)$ **c.** $y = f^3(\theta)$

Fig. 1.9. Periodic points of circle map

First, if a period two point does exist, then $f(\theta_1) = \theta_2$ and $f(\theta_2) = \theta_1$. Combining the terms leads to the arrangement

$$\theta_1 = f(\theta_2) = f(f(\theta_1)),$$

which means that a period two point is the fixed point for $f^2(\theta)$ where f^2 represents the composition $f \circ f$. So, a way to find the period two points is to create a rough graph for $y = f^2(\theta)$ and find whether it has fixed points. That is, we wish to identify all points where the graph of $y = f^2(\theta)$ crosses the $y = x$ diagonal.

Actually it is easy to create a rough graph of $y = f^2(\theta)$. After all, since $f^2(A) = f(f(A)) = f(A \cup B)$, the graph of f^2 over A must resemble the graph of f over $f(A) = A \cup B$; that is, the graph of f over $[0, 2\pi)$. The approach, then, is to squeeze the Fig. 1.9a graph of $y = f(\theta)$ over the interval A: this is done in the first half of Fig. 1.9b. Similarly, since $f^2(B) = f(f(B)) = f(A \cup B)$ where the orientation is preserved, the graph of $y = f^2$ over B squeezes the full Fig. 1.9a graph of f over interval B. Even if the details are not correct, the number of times this rough graph must cross the $y = x$ diagonal proves that there are two new fixed points for $f^2(\theta)$; let them be θ_1 and $\theta_2 = f(\theta_1)$.

Notice how this construction divides the full interval into four subintervals. These subintervals correspond to terms AA, AB, BA, and BB. So, the first f^2 fixed point corresponds to repeating the sequence AB while the second one represents a repeating BA. (Notice the tacit use of the fact that f is continuous and monotonic on each interval; in particular, the f image does not reverse direction as it does with the f over B in Fig. 1.8.)

To find the period 3 points, or the fixed points of $y = f^3(\theta) = f(f(f(\theta)))$, notice that $f^3(A) = f^2(f(A)) = f^2(A \cup B)$. In other words, over the $A = [0, \pi)$ region, squeeze in the full graph of Fig. 1.9b. Similarly, since the mappings are orientation preserving, the graph of $y = f^3(B)$ is a squeezing of the graph of $y = f^2$ from Fig. 1.9b over region B. All of this identifies the Fig. 1.9c graph of $y = f^3$ with its six fixed points. Using the above argument, the eight regions from left to right are

$$AAA,\ AAB,\ ABA,\ ABB,\ BAA,\ BAB,\ BBA,\ BBB$$

so the six period three points are identified, respectively, with the repeated blocks of AAB, ABA, ABB, BAA, BAB, BBA.

The same kind of argument applies to Newton's method and the period-three graph of Fig. 1.8. The main difference is that with Fig. 1.8, the image of the B region is reversed, so the "squeezed graph" must be reversed. Of course, while Newton's method reproduces portions of regions, the existence of periodic points follows from the geometry. Simple arguments; nice conclusions!

1.4.3 The forced Van der Pol equations

All of this material is being introduced in order to describe a problem about the rings of Saturn. But, before doing so, I need to outline a nice argument developed by Mark Levi [36, 37] to analyze the periodically forced Van der Pol equation. While a plausibility argument outlining Levi's arguments suffices for my purposes, the reader is strongly encouraged to read Levi's papers to fully enjoy the details.

The periodically forced Van der Pol equations are given by

$$\epsilon x'' + (x^2 - 1)x' + \epsilon x = b \sin(t), \tag{1.19}$$

where ϵ has a small but fixed value and b is the forcing amplitude: these equations arise from the study of electrical circuits containing that ancient device of vacuum tubes. As shown next, solutions for these equations have a nice and regular "beat." Of course, each of us, if we wish to stay alive, consistently experience another kind of "regular, periodic beat" in our chest. Thus it is interesting but not surprising that Van de Pol used versions of these equations to model and understand the heart's behavior. In fact,

friends of mine from the medical field who study the mathematics of body
organs know the Van der Pol equations strictly from this medical context.

Intuition about the behavior of the Eq. 1.19 system can be obtained
by rewriting these equations in the following form that uses the equality
$\frac{x^3}{3} - x = \int (x^2 - 1)\,dx$. It follows that

$$x' = \frac{1}{\epsilon}(y - [\frac{x^3}{3} - x]), \quad y' = -\epsilon x + b\sin(t).$$

If $b = 0$ and ϵ has a sufficiently small value, then the intuitive sense derived
from the second equation is that the miniscule y' value allows only slow and
minor changes in the y value. On the other hand, the large ϵ^{-1} multiplier for
x' significantly accentuates differences in the first equation. This argument
suggests an active x' change causes the solution to rapidly approach and
remain close to the curve given by $x' = 0$; that is, the solution should
remain near the curve

$$y = \frac{x^3}{3} - x.$$

The graph of this curve is given in Fig. 1.10a.

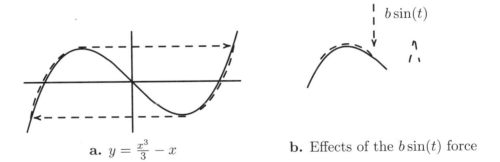

a. $y = \frac{x^3}{3} - x$ **b.** Effects of the $b\sin(t)$ force

Fig. 1.10. The forced Van der Pol equation

This story argues that the solution should follow reasonably close to the
dashed lines where, when sliding over the two precipices given by the local
maximum and minimum of $y = \frac{x^3}{3} - x$, the solution rapidly moves to the
other leg in an essentially horizontal manner. (It is essentially horizontal
because y' has a small value compared to x'.) This argument has been
made precise in different ways.

Now consider the periodic forcing effect of $b\sin(t)$. To develop intuition
for what should happen, Levi considered the effect of the dynamics on line
segments, or strips, of initial conditions. Without the forcing term, this line

segment moves like a well-behaved water snake sliding along the current given by the dashed Fig. 1.10a curve. Something different, however, can occur with the forcing function and a sufficiently large value of b.

At this stage we need a "thought experiment." Suppose, just suppose, that when the strip—that snake—starts to slide over the edge and prepares for a rapid movement to the right as depicted in Fig. 1.10b, the $\sin(t)$ timing allows a large b value to force portions of the strip downwards. The front part of our snake, then, is forced to twist into a "V." However, only a portion of this strip encounters this force; the $\sin(t)$ term reverses sign so that it no longer is forcing the solution downwards. This means that the rest of the strip is not affected when it crosses the threshold. The combined effect is that our "snake" is chasing its tail to create a compressed and rotated "V" that now slides along the solid Fig. 1.10 curve.

The consequences should be fairly clear. Instead of a long strip monotonically making its rounds of the curve, this strip has a kink as represented by the dashed lines to the right of Fig. 1.10. Thus this kinky affect, somewhat resembling the graph in Fig. 1.8, goes around and around to get further kinks representing higher iterates. Consequently, with appropriate care, an analysis similar to that given about Fig. 1.8 applies showing the chaotic effects allowed by this system.

With this approach of following a strip of initial conditions, Levi reduces portions of periodically forced systems to maps from the interval to the interval. In other words, this kinky mathematics captures subtle expansion and recurrence effects of the motion: it identifies potentially complex dynamical behavior.

What is next?

A message to be taken from the above geometric arguments is that it is possible for the combination of *expansion* and *recurrence* of motion to cause surprisingly complex dynamical behavior. This claim is of particular interest to anyone interested in the N-body problem and celestial mechanics. After all, our solar system enjoys plenty of recurrence with the planets making their regular trips about the Sun. Expansion? Well, the first two sections of this chapter showed how expansion and strong forces accompany close approaches of even well-behaved bodies. Maybe this combination will lead to interesting problems. As indicated in the next section, this is the case.

1.5   The rings of Saturn

This chapter started with Galileo's forced recantation of his views: it ends with a problem linked to his wondrous 1610 discovery of the rings of Saturn by use of his telescope.[11] A problem, of course, is to understand why the rings are there. This is a general issue because, as space exploration has proved, rings have been sighted with other planets. While the rings of other planets are not as dramatic, the important point is that they are there. What are the dynamics? One aspect of these dynamical concerns will be addressed in this section; another will be described at the end of the third chapter.

We now know that while the ring system extends quite a distance from Saturn, the thickness is, in fact, quite thin on the astronomical scale: it is only that of a two story building. The particles forming the rings range from dust to objects about 15 feet in diameter. But rather than forming a nice, circular ring, there are portions that resemble the braiding of hair. The research problem proposed here—a problem that is not necessarily easy but it does seem to be doable— is to develop a dynamical explanation.

1.5.1   Kinky behavior

To provide background, in April, 1973, Pioneer 11 was launched on a long journey to visit Jupiter; later it dropped by Saturn to make the first direct observations of this planet (in 1979). One of the Pioneer 11 discoveries was a new ring for Saturn, imaginatively called the "F ring." What made the discovery unusual, as clarified by pictures from a November 1980 visit by Voyager 1, is that the F-ring appeared to be involved in some kind of "kinky" behavior: two of its three strands provide a braided appearance. Whatever the source, this kinkiness appears to be have been short-lived, rather than a permanent phenomenon, as judged by the pictures of Voyager 2 taken less than a year later. At the later time, the F-ring now was more regular with non-intersecting braided strands.

Adding to the mix are the two shepherds of the F-ring, Prometheus and Pandora.[12] Prometheus, the inner moon, is named after brother of Atlas and Epimetheus—he is the one who stole fire from Zeus and gave

---

[11]More precisely, Galileo knew he found something, but, because of the limited power of his equipment, he was not completely sure what were the "ear like" appendages that he observed. Were these moons, or, as Galileo initially thought, two stars circling Saturn? It took a stronger telescope to recognize that they are rings.

[12]Both of these moons of Saturn were discovered in 1980 by S. A. Collins and D. Carlson by carefully examining Voyager 1 photos.

it to mankind. Pandora, the outer moon, is named after the first woman who was created by Zeus to punish man for Prometheus' stealing of fire—the punishment consisted of the evils released once Pandora opened her "Pandora's box." The astronomical situation, not the mythological one, is beautifully displayed in one of Voyager's photos given by Fig. 1.11.

Fig. 1.11. Prometheus, F-ring, and Pandora;
this picture is used thanks to the courtesy of NASA/JPL-Caltech.

The question is whether the motions of these two moons are responsible for the braiding of the F-ring. This conjecture is so natural that I must assume it has been advanced by many people, yet I do not know of anyone who has provided a mathematical verification of this suggestion. On the other hand, by pulling together all that has been discussed in this chapter, it is possible to propose a natural model—and a mathematical approach—to explain these braids. Let me outline the notions.

1.5.2   A model

The first step is to compute a portion of the obit of Pandora relative to the position of Prometheus (i.e., put the system in a rotating coordinate system based on the motion of Prometheus). According to what we determined earlier (page 5) when discussing the orbit of Mars relative to the Earth, we know that in this system Pandora's orbit must exhibit a loop. Even more: the size of the loop can be slightly enhanced because, although the eccentricity of Pandora's orbit, 0.0042, is small, it is about twice that of Prometheus's eccentricity of 0.0024. By using the earlier argument about the apparent location of the Sun as viewed from Mercury, it can be shown that the loop size of Pandora's apparent orbit can be slightly expanded.

Now consider a dust particle circling Saturn between Prometheus and Pandora. Just because the particle goes about Saturn, its motion has a sense of "recurrence." But, should that particle be somewhere near Pandora when her looping occurs, this proximity creates a stronger pull on the dust—it generates an expansion effect. In other words, the F-ring dynamics, as modified by the presence of Prometheus and Pandora, provides a setting of "expansion and recurrence." This setting is precisely what is needed to suggest the complicated dynamics discussed earlier.

The situation is depicted in Fig. 1.12 where the dot on the left locates the position of Prometheus, the curved line corresponds to a strip of particles in the F-ring, and the loop represents the close incursion of Pandora. To understand the effects of the loop, borrow Levi's approach of using a moving strip—his snake traveling through the region—and consider what happens to various parts of its body. To do so, start with Pandora on the upper part of the loop not yet near its closest approach and where a small F-ring strip is not quite symmetrically centered on the figure because its head is on the circle directly between Pandora and Prometheus. For purposes of argument, assume that both particles in the F-ring and Pandora are moving in a general counter-clockwise direction. (This assumption just corresponds to whether we are looking at the ring from the North or South pole of Saturn.) According to Kepler's second law (Eq. 1.10), particles in the F-ring are moving faster than Pandora.

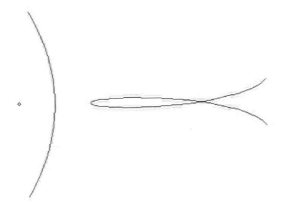

Fig. 1.12. Prometheus, F-ring, and Pandora in a rotating system

The "thought-experiment" shows that with this orientation, Pandora pulls harder on the snake's head than tail. As the snake slithers past, Pandora reaches the bottom of the loop and now exerts an even stronger pull on the snake's midsection. With its faster motion, the snake moves on, while Pandora is receding in the opposite direction causing the tail to be influenced more by Prometheus than Pandora.

What we have, then, is that various portions of the snake are pulled toward Pandora, and even pulled backwards when the relative motion of Pandora is going backwards. While I know nothing about knitting, this descriptions of how the strip is being pulled and pushed around sounds much like the "Knit one, purl two" phrases uttered by my grandmother.

Now, if two braids that normally would be well behaved run into this looping effect, the differences in speed (again, according to Kepler's law) would require the strips on the braids to encounter this difference in gravitational pull at changing times; we must expect that they would cross and look like a braid. Can this (weak) plausibility argument be made into a more rigorous argument? I don't know, but all the elements suggesting something like this must occur are in place. To continue, notice how this argument suggests that when Prometheus and Pandora are sufficiently separated—a setting where we will not expect this kinky looping behavior to occur— then we must expect the appearance of the F-ring to be more standard: this is consistent with what Voyager 2 reported.

The challenge is given: the technical difficulty derives from the added degrees of freedom that are above that of the Levi example. Yet, I expect that this is the kind of problem that can be solved.

Chapter 2

Central configurations

Building on our description of Saturn's kinky F-ring, given at the end of Chap. 1, it is worth wondering about the general dynamical behavior of the rings of Saturn. These rings cannot be dismissed as representing transient phenomenon—after all, they have survived for who knows how long—and we now know that other planets have their own rings. Therefore it is reasonable to speculate that they are stable. Are they? If so, why? If not, why not?

Very little is known about this question. One fact is clear: the natural combination of the mathematical challenge and astronomical interest make this topic a rich and attractive research issue. Some comments, along with new research problems describing new twists about this issue are in the last section of the next chapter.

The mathematical study of Saturn's rings goes back at least to 1856 when Maxwell[1] developed a model for their dynamics based on what are called *central configurations* and *relative equilibrium* orbits. For the moment, consider these relative equilibrium orbits to be circular orbits that retain the same configuration—these configurations will be the "central configurations"—for all time. As we will discover next, these central configurations play a surprisingly important role in the understanding of the Newtonian N-body problem while posing another set of important but difficult research issues.

---

[1] James Clark Maxwell (1831-1879) received his 1854 degree from Trinity College, Cambridge, where he met, and competed with Edward Routh on the mathematical tripos (Routh won) and the Smith prize (Maxwell won). Small world. In the history of mathematics, Routh is known as an exceptional teacher, but readers of this book probably know him better through his work on linear approaches toward stability and the Routh–Hurwitz test.

2.1   Equations of motion and integrals

First we need the equations of motion for the N-body problem. As indicated in Fig. 2.1, let m_j and $\mathbf{r}_j$, $j = 1, \ldots, N$, be, respectively, the mass and the position vector of the jth particle in an inertial coordinate system.[2] Using Newton's "mass times acceleration equals force," the equations of motion are

$$m_j \mathbf{r}_j'' = \mathbf{F}_j$$

where $\mathbf{F}_j$ is the force exerted on $\mathbf{r}_j$ by the other particles and $\mathbf{r}_j''$ is the acceleration.

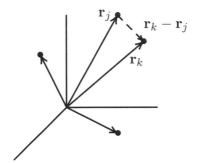

Fig. 2.1. Particle positions

The Newtonian inverse square force between two bodies is $\frac{G m_j m_k}{r_{jk}^2}$ where $r_{jk} = |\mathbf{r}_j - \mathbf{r}_k|$ and G is the gravitational constant that, by appropriately choosing dimensions, we can conveniently ignore by setting it equal to unity. As depicted in Fig. 2.1, the kth particle pulls the jth toward it in the direction $\frac{\mathbf{r}_k - \mathbf{r}_j}{r_{jk}}$, so, by summing the forces contributed by all of the bodies, the equations of motion for $\mathbf{r}_j$ become

$$m_j \mathbf{r}_j'' = \sum_{j \neq k} \frac{m_j m_k (\mathbf{r}_k - \mathbf{r}_j)}{r_{jk}^3}, \quad j = 1, \ldots, N. \tag{2.1}$$

Exploiting the enormous symmetry afforded by Eqs. 2.1 leads to nine of the integrals of motion. For instance, by summing this equation over all $j = 1, \ldots, N$, each $\frac{m_j m_k (\mathbf{r}_k - \mathbf{r}_j)}{r_{kj}^3}$ term on the right-hand side is accompanied

---

[2]For us, an inertial coordinate system is one where the equations of motion hold.

by a cancelling $\frac{m_k m_j (\mathbf{r}_j - \mathbf{r}_k)}{r_{jk}^3}$ anti-symmetric term leading to

$$\sum_{j=1}^{N} m_j \mathbf{r}_j'' = \mathbf{0}.$$

Integrating twice yields the integrals of motion

$$\sum_{j=1}^{N} m_j \mathbf{v}_j = \mathbf{A}, \quad \sum_{j=1}^{N} m_j \mathbf{r}_j = \mathbf{A}t + \mathbf{B}$$

where $\mathbf{v}_j = \mathbf{r}_j'$ and $\mathbf{A}, \mathbf{B}$ are vector constants of integration. A standard choice is to select $\mathbf{A} = \mathbf{B} = \mathbf{0}$. The reason is that, by letting $M = \sum_{j=1}^{N} m_j$ represent the total mass, these six integrals become

$$\frac{1}{M} \sum_{j=1}^{N} m_j \mathbf{v}_j = \frac{1}{M} \sum_{j=1}^{N} m_j \mathbf{r}_j = \mathbf{0}, \qquad (2.2)$$

which assert that the center of mass of the system is fixed at the origin of the inertial coordinate system. In other words, selecting $\mathbf{A} = \mathbf{B} = \mathbf{0}$ places the emphasis on the relative motion of the N particles.

Three more integrals, which reflect the rotational symmetry, are nicely captured by using the vector cross product. Take the vector product $\mathbf{r}_j \times \mathbf{r}_j''$ in Eq. 2.1 and sum over all j. In the double summation, each $\mathbf{r}_j \times \mathbf{r}_j = \mathbf{0}$ and each $\frac{m_j m_k \mathbf{r}_j \times \mathbf{r}_k}{r_{jk}^3}$ term is paired with a cancelling $\frac{m_k m_j \mathbf{r}_k \times \mathbf{r}_j}{r_{kj}^3}$ resulting in

$$\sum_{j=1}^{N} m_j (\mathbf{r}_j \times \mathbf{r}_j'') = \frac{d}{dt} [\sum_{j=1}^{N} m_j (\mathbf{r}_j \times \mathbf{v}_j)] = \mathbf{0}.$$

Integration leads to the integral of angular momentum

$$\sum_{j=1}^{N} m_j \mathbf{r}_j \times \mathbf{v}_j = \mathbf{c} \qquad (2.3)$$

where $\mathbf{c}$ is a vector constant of integration that normally is chosen to be $\mathbf{c} = (0, 0, c)$.

This choice of $\mathbf{c}$ makes the x–y plane the *invariable plane:* it is the plane of motion for the N-body coplanar problem. To see why this is so, if all particles are in the same plane, then $\mathbf{r}_j \cdot [\sum_{j=1}^{N} m_j \mathbf{r}_j \times \mathbf{v}_j] = 0$, or $\mathbf{r}_j \cdot \mathbf{c} = 0$ for each j. Thus, the plane of motion is defined by the normal vector $\mathbf{c}$.

The remaining integral uses the *self-potential* (the negative of the potential energy)

$$U = \sum_{j<k} \frac{m_j m_k}{r_{jk}}, \tag{2.4}$$

which allows the equations of motion to expressed as

$$m_j \mathbf{r}_j'' = \frac{\partial U}{\partial \mathbf{r}_j} = (\frac{\partial U}{\partial x_j}, \frac{\partial U}{\partial y_j}, \frac{\partial U}{\partial z_j}), \quad j = 1, \dots, N. \tag{2.5}$$

By taking the scalar product of both sides of Eq. 2.5 with respect to $\mathbf{r}_j'$, summing over all j, and using standard rules from calculus, we obtain

$$\frac{d}{dt}[\frac{1}{2} \sum_{j=1}^{N} m_j \mathbf{v}_j^2] = \sum_{j=1}^{N} m_j (\mathbf{r}_j' \cdot \mathbf{r}_j'') = \sum_{j=1}^{N} \frac{\partial U}{\partial \mathbf{r}_j} \cdot \mathbf{r}_j' = \frac{d}{dt} U.$$

Integrating the extreme ends of this expression yields the last integral, the *conservation of energy integral,*

$$T := \frac{1}{2} \sum_{j=1}^{N} m_j v_j^2 = U + h \tag{2.6}$$

where h—the total energy—is a constant of integration and T is the kinetic energy.

This is it: these are the only known integrals. With only ten of them, it dooms all hope one might harbor about trying to solve the N-body problem via the implicit function theorem.[3] After all, with $6N$ variables and only ten equations we are in debt to the level of $6N - 10$ integrals. So, rather than trying to solve the N-body problem, an alternative approach is to examine special solutions and particular features. A natural starting place is with "central configurations."

2.2   Central Configurations

Just imagine how much simpler it would be if we could replace the complicated $\mathbf{r}_j'' = \frac{1}{m_j} \frac{\partial U}{\partial \mathbf{r}_j}$ equations with the much easier system

$$\mathbf{r}_j'' = \lambda \mathbf{r}_j, \quad j = 1, \dots, N. \tag{2.7}$$

---

[3]We sometimes forget that, for a long time, this strategy of finding enough integrals was an expected approach to solve the N-body problem. In a 1893 talk to the New York Mathematical Society, for instance, Simon Newcomb [54] (see Saari [91])—a founding member and first president of the American Astronomical Society and the fourth president of the American Mathematical Society—lamented that, "To the problem of three bodies we have not added one of the integrals necessary to the complete solution."

If, for instance, λ is a fixed, negative constant independent of the particle j, then one solution would be

$$\mathbf{r}_j(t) = A_j(\cos(\sqrt{|\lambda|}t), \sin(\sqrt{|\lambda|}t), 0), \quad j = 1, \ldots, N.$$

Aristotle would have been delighted with this N-body motion because the particles rotate in an uniform circular motion in the invariable plane.

Interestingly, for any choice of masses, choices of configurations and velocities can be found that allow this Eq. 2.7 replacement. (The search for these special configurations is in Chap. 3: in this chapter, I describe the importance of these configurations.) A quick way to identify these special N-body settings is to equate Eqs. 2.5, 2.7. By doing so, it follows that these configurations must satisfy the condition $\lambda \mathbf{r}_j - \frac{1}{m_j}\frac{\partial U}{\partial \mathbf{r}_j}$ for each j.

Definition 2.1 *The N-particles form a central configuration at time t if there exists a scalar λ so that*

negative? √ p.40

$$\lambda m_j \mathbf{r}_j = \frac{\partial U}{\partial \mathbf{r}_j}, \quad j = 1, \ldots, N. \tag{2.8}$$

In other words, a central configuration occurs whenever the particles line up in the particular fashion where each particle's position vector is the same scalar multiple of the particle's acceleration vector. This description makes central configurations sound like a rare and unlikely phenomenon. Indeed, if these configurations described only rare, particularly well-behaved N-body solutions, we would dismiss them as being of amusing passing interest. But this is not the case: central configurations play a "central" role in understanding N-body behavior. A theme of this chapter is to indicate why this is so.

As we will learn, these configurations tend to arise whenever the same configuration is rigidly preserved during the motion, or whenever the N-body gravitational pressures force a "balancing" among the limiting behaviors of particles. As outlined at the end of the next chapter, for instance, central configurations even occur when modeling the rings of Saturn. In Chap. 4, I show that this balancing act always occurs when particles collide. Even more (Saari [90]), these configurations arise with galaxies separating from one another as time goes to infinity. Consequently, to fully appreciate the dynamical behavior of collisions and expansions, we need to better understand central configurations and their properties.

In a different direction, when S. Smale [102] analyzed the topology of the manifolds defined by fixed values of the angular momentum and energy integrals (Eqs. 2.3, 2.6), he discovered that the central configurations play a key role in identifying bifurcation behavior for the topological description. (Around this time R. Easton [19] carried out a similar analysis for the three-body problem.) The same phenomenon occurs when examining the restrictions on the permitted geometric configurations (Marchal and Saari [41]). The reason for the ubiquity of these configurations is that in both the topological and geometric studies, the limiting behavior is characterized by an orbit where the configuration remains fixed: thus we must expect a central configuration. These examples make it clear that central configurations play a "central" role in the N-body problem.

2.2.1   Why central configurations are important

A way to develop intuition why we should expect central configurations to accompany the above settings is to play a "suppose that" game.

Collisions

For a first thought experiment, "suppose that" when particles approach a collision at the origin at time $t = 0$, they do so in a

$$\mathbf{r}_j = \mathbf{A}_j t^\beta + o(t^\beta), \text{ as } t \to 0, \, j = 1, \ldots, N, \qquad (2.9)$$

asymptotic manner where $\mathbf{A}_j$ is an unknown vector constant, exponent β is an unknown constant, and $o(t^\beta)$ represents higher order terms that approach zero faster than t^β.

Next, "suppose that" we can differentiate the asymptotic relationship in Eq. 2.9. (At times it is legitimate to differentiate this relationship, but, as described in Sect. 4.5.2 (page 173), there exist "oscillating" situations—precisely the type that might occur in celestial mechanics—where this cannot be done.) Whenever differentiation is permitted, we have that

$$\mathbf{r}_j'' = \beta(\beta - 1)\mathbf{A}_j t^{\beta-2} + o(t^{\beta-2}).$$

Substituting the terms into the equations of motion Eq. 2.1 leads to the asymptotic expression

$$\beta(\beta - 1)m_j \mathbf{A}_j t^{\beta-2} = \sum_{k \neq j} \frac{m_j m_k (\mathbf{A}_k - \mathbf{A}_j)}{|\mathbf{A}_k - \mathbf{A}_j|^3} t^{-2\beta} + o(t^{-2\beta} + t^{\beta-2}) \qquad (2.10)$$

for each $j = 1, \ldots, N$.

This "suppose that" scenario describes an equality in the first terms of this asymptotic relationship where the lead exponents on t on both sides must agree; that is,

$$\beta - 2 = -2\beta, \text{ or } \beta = \frac{2}{3}.$$

Should this story have any validity, collisions asymptotically behave like

$$\mathbf{r}_j \sim \mathbf{A}_j t^{2/3} \text{ as } t \to 0. \tag{2.11}$$

Moreover, by substitution, the lead $\mathbf{A}_j$ terms satisfy the equations

$$-\frac{2}{9} m_j \mathbf{A}_j = \sum_{k \neq j} \frac{m_j m_k (\mathbf{A}_k - \mathbf{A}_j)}{|\mathbf{A}_k - \mathbf{A}_j|^3} \text{ for } j = 1, \dots, N. \tag{2.12}$$

Because Eq. 2.12 agrees with the defining expression for central configurations (Eq. 2.8), we must anticipate that, asymptotically, collisions form central configurations as the particles tend toward each other. When discussing collisions, we will show that this intuition is correct.

Asymptotic growth as time goes to infinity

In the asymptotic expansion of the Newtonian N-body problem (Saari [75, 90], Marchal and Saari [40]), one motion describes the separation of the center of masses of, say, groupings of particles that are relatively bounded. To use suggestive terminology, call these groupings of relatively bounded particles "galaxies." It turns out that these 'galaxies" are clustered: they separate from one another like t^β where $\beta < 1$.

To examine this motion, let $\boldsymbol{\rho}_j(t)$ be the position of the jth galaxy of this cluster relative to the center of mass of the cluster. Next, "suppose that"

$$\boldsymbol{\rho}_j \sim \mathbf{A}_j t^\beta \text{ as } t \to \infty.$$

Assume that this asymptotic relationship can be differentiated. (This assumption is more problematic than the one made for collisions.) Substituting these relationships into the equations of motion for

$$\boldsymbol{\rho}_j = \frac{1}{M_j} \sum m_k \mathbf{r}_k,$$

where the summation is over all particles in the jth galaxy and M_j is the total mass of this galaxy, we have that

$$\beta(\beta - 1) M_j \mathbf{A}_j t^{\beta-2} = \sum_{k \neq j} \frac{M_j M_k (\mathbf{A}_k - \mathbf{A}_j)}{|\mathbf{A}_k - \mathbf{A}_j|^3} t^{-2\beta} + o(t^{-2\beta} + t^{\beta-2})$$

where the error terms reflect interactions of particles within each galaxy and among clusters of galaxies.

Using the same argument as for collisions, it follows that if the separation of the galaxies has the indicated expansion and if the asymptotic relations can be differentiated, then $\beta = \frac{2}{3}$ and the separating galaxies must, as $t \to \infty$, asymptotically define the central configuration equation

$$-\frac{2}{9} M_j \mathbf{A}_j = \sum_{k \neq j} \frac{M_j M_k (\mathbf{A}_k - \mathbf{A}_j)}{|\mathbf{A}_k - \mathbf{A}_j|^3}$$

for each galaxy in the cluster.

As verified in (Saari [90]), many of these "suppose that" descriptions can be validated. In these settings, then, we have a nice description of the evolution of the universe. On the other hand, because the N-body problem is precisely the academic topic where "chaos" was originally discovered (the references on page 137 describe this "chaos story"), we must anticipate surprises to arise. They do.

Relative equilibria

Of particular interest for this and the next chapter is the special motion where the bodies rotate in circular motion while *keeping the same configuration for all time*. As already advertised, these are known as *relative equilibria*.

Expressing this behavior in terms of complex variables, the motion of the jth particle must be

$$z_j(t) = e^{i\omega t} z_j(0)$$

where ω is a fixed constant. With this expression, there are no concerns or challenges about differentiation, so

$$z_j''(t) = -\omega^2 e^{i\omega t} z_j(0).$$

Substituting this last expression into the equations of motion leads to

$$
\begin{aligned}
m_j z_j'' = -m_j \omega^2 e^{i\omega t} z_j(0) &= \sum_{k \neq j} \frac{m_k m_j (z_k(t) - z_j(t))}{|z_k(t) - z_j(t)|^3} \\
&= e^{i\omega t} \sum_{k \neq j} \frac{m_k m_j (z_k(0) - z_j(0))}{|z_k(0) - z_j(0)|^3}.
\end{aligned}
\tag{2.13}
$$

With the cancellation of the $e^{i\omega t}$ term from both sides of this equation, it follows that if the particles rotate in circular motion while retaining the same configuration for all time—a relative equilibria—the particles satisfy

$$-m_j \omega^2 z_j(0) = \sum_{k \neq j} \frac{m_k m_j (z_k(0) - z_j(0))}{|z_k(0) - z_j(0)|^3} \quad \text{for all } j, \tag{2.14}$$

which means that they must form a central configuration.

Bifurcation of energy-angular momentum surfaces

Even a plausibility argument explaining why central configurations arise when the energy-angular momentum surfaces have a topological bifurcation is too long for our purposes. (A description is in Saari [90].)

To roughly describe what happens, the angular momentum and energy integrals define a manifold characterizing the motion admitted by these constants: they indicate what kinds of constraints are imposed on the motion. It turns out that there are critical values of these integrals that cause bifurcations in the structure of these manifolds. Even more: for the coplanar N-body problem, the bifurcation region is occupied by an orbit where, with uniform circular motion, the particles retain the same configuration—as we should begin to suspect, this configuration is a central configuration. Thus the topological bifurcations for the coplanar problem are accompanied by central configurations. In contrast, a bifurcation in a three-dimensional physical spaces (described in [90]) can occur where it is impossible to sustain an orbit with a fixed configuration: here we must expect (and this happens) that the bifurcations define different kinds of configurations.

Terminology

To complete this introduction to central configurations, let me introduce standard terminology coming from Wintner [112]. *Homographic motion* is where a configuration is preserved for all time. Nothing is stated in the definition about the configuration, but the discussion starting in Sect. 2.3.4 shows for the Newtonian N-body problem that this preserved configuration is a central configuration. The "Newtonian N-body" qualifier is important because this assertion fails to hold for some force laws. While the conclusions do hold at least for the inverse p force laws for $1 < p < 3$, where the Newtonian force has $p = 2$, strange things happen for $p = 3$.

A special type of homographic solutions is where the solutions behave like a rotating rigid body; i.e., the distances among all of the bodies remain fixed so the particles rotate without change in scale. Such a solution, which was discussed above, is called a *relative equilibrium solution*.

The other extreme is where the solution allows the particles to change scale size while preserving the configuration but with no rotation of any kind. This means that each particle is constrained to lie on a straight line passing through the center of mass of the system. Such a solution is called *homothetic.* According to the definition of the angular momentum integral (Eq. 2.3), homothetic behavior can occur only if $c = 0$.

2.2.2   Value of λ

Now that we have an intuitive appreciation as to why central configurations are central to the study of the N-body problem, it is time to examine the actual configurations. A natural first question is to understand what they look like. To build the background needed to address this difficult question, I start with some simpler, more basic concerns. What is the value of the scalar λ that is introduced in Eq. 2.8? How are central configurations found?

   A handle on these questions is obtained by using one-half the polar moment of inertia defined as

$$I = \frac{1}{2} \sum_{j=1}^{N} m_j \mathbf{r}_j^2. \tag{2.15}$$

By use of I, the λ value can be computed and the central configurations can be shown to be the critical points of a particular function that I call the *configurational measure.*

Theorem 2.1 *For a central configuration,*

$$\lambda = -\frac{U}{2I}. \tag{2.16}$$

A configuration $\mathbf{R} = (\mathbf{r}_1, \ldots, \mathbf{r}_N)$ *is a central configuration if and only if*

$$\nabla I U^2(\mathbf{R}) = \mathbf{0}. \tag{2.17}$$

Proof: Taking the dot product of both sides of Eq. 2.8 with $\mathbf{r}_j$ and then summing over all j leads to

$$2\lambda I = \lambda \sum_{j=1}^{N} m_j(\mathbf{r}_j \cdot \mathbf{r}_j) = \sum_{j=1}^{N} \frac{\partial U}{\partial \mathbf{r}_j} \cdot \mathbf{r}_j. \tag{2.18}$$

To simplify the last term, recall from Euler's Theorem that if a differentiable function is homogeneous of degree β (i.e., $f(tx_1, \ldots, tx_N) = t^\beta f(x_1, \ldots, x_N)$ for all $t > 0$), then it satisfies

$$\sum_{j=1}^{N} \frac{\partial f}{\partial x_j} x_j = \beta f.$$

(For a proof, differentiate $f(tx_1, \ldots, tx_N) = t^\beta f(x_1, \ldots, x_N)$ with respect to t and evaluate at $t = 1$.) Because U is homogeneous of degree -1, Eq. 2.18 becomes $2\lambda I = -U$, and Eq. 2.16 follows.

To prove Eq. 2.17, because $m_j \mathbf{r}_j = \frac{\partial I}{\partial \mathbf{r}_j}$, we have that Eq. 2.8 is true if and only if $\lambda \nabla I = \nabla U$. By substituting the established value of λ into the equation, this expression holds if and only if $U \nabla I + 2I \nabla U = 0$. Multiplying by U and using the product rule shows that this expression holds if and only if $\nabla I U^2 = \mathbf{0}$. $\square$

Other potentials

As an aside, it is worth pointing out that the same argument holds for a variety of potentials. For instance, the inverse p force law, $p \neq 1,$[4] defines the self-potential

$$U_p = \frac{1}{p-1} \sum_{j<k} \frac{m_j m_k}{r_{jk}^{p-1}} \tag{2.19}$$

where $p = 2$ is the Newtonian potential of Eq. 2.4. As U_p is homogeneous of degree $1 - p$, the corresponding value of λ for Eq. 2.16 becomes

$$\lambda_p = (\frac{1-p}{2}) \frac{U}{I}.$$

The new version of Eq. 2.17 asserts that a central configuration corresponds to the critical point

$$\nabla I U^{2/(p-1)} = \mathbf{0} \text{ or, equivalently, } \nabla I^{(p-1)/2} U = \mathbf{0}. \tag{2.20}$$

It is interesting to apply the Sect. 2.2.1 "suppose that" stories with these other force laws. By doing so, we learn to anticipate that collisions, or separating galaxies, will asymptotically approach a central configuration where the $\beta = \frac{2}{3}$ exponent is replaced with $\beta = \frac{2}{p+1}$: this is what happens.

Configurational Measure

As another "aside," because IU^2 is homogeneous of degree zero, its value does not depend upon the size of the configuration defined by the particles, but only on its shape. In Chap. 3 we will see for the three body problem that the configurational measure has its minimum value at an equilateral triangle. However, when two particles approach each other while the third is

---

[4]For $p = 1$, use $U = \sum m_j m_k \ln(r_{jk})$. This expression arises in the analysis of vortices.

a reasonable distance away, the increase in the U value requires IU^2 to have an arbitrarily large value: thus the configurational measure values define an interval $[d, \infty)$. Treat IU^2, or the $\sqrt{I}U$ value, as a "configurational measure" that tells us something about the shape of the configurations. Restating Thm. 2.1 with this terminology leads to the assertion that *a central configuration is a critical point of the configurational measure.*

2.2.3   Equivalence classes of configurations

The next step is to use an expression of I that uses the relative distances between particles.

Proposition 2.1 *The term I can be expressed as*

$$I = \frac{1}{2M} \sum_{j<k} m_j m_k (\mathbf{r}_j - \mathbf{r}_k)^2 = \frac{1}{2M} \sum_{j<k} m_j m_k r_{jk}^2 \qquad (2.21)$$

where M is the total mass.

Proof: By expansion,

$$\begin{aligned}
\tfrac{1}{2M} \sum_{j<k} m_j m_k (\mathbf{r}_j - \mathbf{r}_k)^2 &= \tfrac{1}{4M} \sum_{j=1}^N \sum_{k=1}^N m_j m_k (\mathbf{r}_j - \mathbf{r}_k)^2 \\
&= \tfrac{1}{4M} \sum_j \sum_k m_j m_k [r_j^2 - 2\mathbf{r}_j \cdot \mathbf{r}_k + r_k^2].
\end{aligned}$$

But $\sum_k \sum_j m_j m_k r_j^2 = \sum_k m_k \sum_j m_j r_j^2 = 2MI$. Similarly $\sum_j \sum_k m_k m_j r_k^2 = 2MI$. Thus, it remains to show that

$$\sum_j \sum_k m_j m_k \mathbf{r}_j \cdot \mathbf{r}_k = \left(\sum_j m_j \mathbf{r}_j\right) \cdot \left(\sum_k m_k \mathbf{r}_k\right) = \mathbf{0}.$$

This equality follows from the center of mass integral Eq. 2.2. □

Notice how Prop. 2.1 allows $\sqrt{I}$ to be viewed as measuring the diameter of the universe while U^{-1} measures the minimum spacing between particles. Consequently, the configurational measure $\sqrt{I}U$ is a particular ratio of measures of the maximum and minimum spacing between particles.

Corollary 2.1 *For the N-body problem, there exist positive constants A, B, C, D, which depend only on the masses, for which*

$$A\sqrt{I} < \max_{j,k} r_{jk} < B\sqrt{I} \qquad (2.22)$$

and

$$C U^{-1} < \min_{j \neq k} r_{jk} < D U^{-1}. \qquad (2.23)$$

Proof: It follows immediately from Eq. 2.21 that

$$I \leq [\frac{1}{2M} \sum_{j<k} m_j m_k](\max_{j,k} r_{jk})^2.$$

Also, if m_0 is the smallest mass value, then, because at each instant of time some choice of r_{jk} is the maximum spacing, we have that

$$I > \frac{m_0^2}{2M}(\max_{j,k} r_{jk})^2.$$

Eq. 2.22 follows immediately. Eq. 2.23 uses the same approach with the definition of U (Eq. 2.4). $\square$

Because this Prop. 2.1 representation of I allows IU^2 to be defined strictly in terms of the mutual distances between particles, it follows that the configurational measure IU^2 is invariant with respect to rotations. Thus if $\mathbf{R}^* = (\mathbf{r}_1^*, \ldots, \mathbf{r}_N^*)$ is a central configuration, then so are all rigid body rotations of $\mathbf{R}^*$. Similarly, as IU^2 is homogeneous of degree zero, if $\mathbf{R}^*$ is a central configuration, then so is $t\mathbf{R}^*$ for any scalar $t \neq 0$. These observations can be expressed in the following manner.

Proposition 2.2 *If $\mathbf{R}^*$ is a central configuration, then so are all configurations in its Euler similarity class.*

When describing and counting central configurations, the discussion is in terms of the defined Euler similarity classes.

Other potentials and even cracks

Incidentally, Prop. 2.2 does *not* hold for all potentials. In the literature, for instance, there are several natural potentials of the form

$$U_{p,q} = U_p + U_q, \quad p \neq q.$$

These choices include the so-called Manev potentials studied by Diacu [15], Perez, and others. (For instance, see Delgado, Diacu, Lacomba, Mingarelli, Mioc, Perez, and Stoica [12].)

A more immediate and familiar example uses the earlier discussion (page 6) about how Einstein explained the bothersome mystery about the 43" of arc advance of the perihelion of Mercury in terms of relativity. Rather than using the actual relativity theory, it appears that Einstein approximated this theory with the $\frac{-1}{r^2} + \frac{\epsilon}{r^4}$ force law where the constant ϵ has a small value of the order of the inverse square of the speed of light.[5] His use of an approximation leads to natural concerns. After all, because the Sun and the planets are not perfect spheres, an approximation in the invariable plane to handle the oblateness effects uses forces of the $\frac{-1}{r^2} + \frac{A}{r^4}$ type.

As another illustration, the potential for the well known Lennard-Jones force law captures a $\frac{A}{r^6} - \frac{B}{r^{12}}$ effect where the first term attracts particles while the second repelling term become dominant when particles come too close to each other. A purpose of the repelling term, then, is to prevent collisions.

The equilibrium configurations of the Lennard-Jones potential help us understand cracks in glass and other objects.[6] To explain, the equilibrium central configurations depend in an interesting manner on the value of N. If the value of N is sufficiently large so that we do not have to worry about boundary effects (so, we could place the N-particles on a torus or a sphere), one equilibrium configuration has the particles evenly spaced in a lattice grid of the type indicated in Fig. 2.2a. With enough particles and/or by ignoring slight differences that might arise on the boundary, we must expect something similar to Fig. 2.2a.

a. Undisturbed   **b.** All cracked up

Fig. 2.2. Equilibrium configurations

Now remove one of the boundary particles, or separate a couple of them to resemble the annoying situation that happens when you get a chip in the windshield of your car, on a valuable piece of crystal, or a slight crack in a windowpane. Because equilibrium configurations require a balance of forces, if the separated particles are far enough apart to reduce the "near-neighbor" attraction effect, we must expect that the new equilibrium situation separates some particles to create a setting with a gap or separation resembling a "crack" as indicated in Fig. 2.2b. (This is particularly so when the constants for the integrals vary due to changes in temperature, etc.) The new equilibria, then, could define "cracks." While this approach provides insight, other

---

[5]On page 88 of the original edition of Pollard's book [60], he provides a simple, elegant description of how this perturbation effect captures the 43" of arc.

[6]These comments are based on my unpublished research in the late 1970s and early 1980s while a consultant at the National Bureau of Standards—now known as the National Institute for Science and Technology.

types of mathematical techniques are traditionally used to study cracks. As such, other than earlier preliminary results or numerical simulations, I doubt whether this analysis has been carefully examined.

Returning to central configurations, because $U_{p,q}$ is *not* homogeneous, it follows immediately that while the central configurations for $U_{p,q}$ are invariant with respect to rigid body rotations, not all of them are invariant with respect to scale. (There are exceptions: the highly symmetric central configurations that occur for all U_p, such as an equilateral triangle or an equilateral tetrahedron, are central configurations for $U_{p,q}$ that are invariant with scalar changes.) This observation makes it easy to show the existence of a continuum of $U_{p,q}$ central configurations; e.g., a continuum of similarity classes of central configurations where the configurations change with the scale size. Using the argument given later, for instance, it follows that the collinear three-body central configurations with these mixed potentials change with scale: the change reflects which potential term dominates.

Newtonian potential

While this continuum of configurations for mixed potentials cause different and interesting dynamical behavior (at collisions, expansions, and even at the topological bifurcations; e.g., U_p may dominate near collisions so this force law dictates the rate of approach while U_q may be the dominate term for expansions), our interest centers almost totally on the Newtonian homogeneous potential. A natural first question is whether there can be a continuum of central configurations in the Newtonian N-body problem. Important question, but we do not know the answer.

As it will be shown in Chap. 3, the Newtonian potential permits precisely the four three-body central configurations that are described in the next chapter. As a preview, there are three collinear configurations: one for each choice of which particle is the middle. The only non-coplanar configuration is an equilateral triangle.

While we would like to determine all central configurations for $N \geq 4$, this turns out to be a surprisingly difficult problem. Indeed, the problems are sufficiently deep, complex, and important that it would be viewed as a major contribution just to establish that there are only a finite number of central configurations for each $N \geq 4$ and all positive masses. This problem, which has frustrated mathematicians for over a century, is described in various places including Winter's classic [112][7] and promoted in Smale's [103] list of

---

[7]This book by Aurel Winter (1903-1958) has had a profound influence on the development of celestial mechanics. The references and cataloguing of "who did what and when"

problems for the twenty-first century.

Progress has been made; e.g., A. Albouby [1] found all four-body central configurations for equal masses: a square, an equilateral triangle with the fourth particle at the center, and an isosceles triangle with the fourth particle on the angle bisector for the one non-equal angle. Later R. Moeckel [49] established that, with the possible exception of a lower dimensional algebraic set of values for the masses, there are only a finite number of four-body central configurations. Then, as the manuscript for this book was being completed, Marshall Hampton and Richard Moeckel [24] announced a computer assisted proof that solves the problem for $N = 4$.

Degenerate central configurations

Before moving on, if there is a continuum of central configurations, it is easy to show that there is a smooth curve $\alpha(t)$ so that $\nabla IU^2(\alpha(t)) = \mathbf{0}$. Of course, $\alpha(t)$ cannot stay in the same Euler equivalence class, so we can assume that it does not change the scale or orientation of the configuration. This structure means that $\alpha'(0)$ is in the kernel of $D(\nabla IU^2)$; namely, in this setting the kernel can consist of more than the rotations and scalar changes of the central configuration. This leads to the following definition.

Definition 2.2 *A central configuration* $\mathbf{A}$ *is said to be non-degenerate if the kernel of* $D_\mathbf{A}(\nabla IU^2)$ *is spanned by directions that correspond to scalar changes of the configuration and rigid body rotations of the configuration. If the kernel is larger, then the configuration is said to be degenerate.*

If there is a continuum of central configurations, they consist of degenerate configurations. However, we know from Palmore [55] that there exist

---

and his systematic description of the problems in mathematical terms may be of more value than his mathematical arguments. Wintner dedicates the book to his Ph.D. advisor Leon Lichtenstein who encouraged him already during his graduate student days to write a book on the three-body problem. It may seem to be premature to advise a graduate student to write such a book! To put this advice into perspective, the hyper-inflation of the 1920s made it difficult for Wintner to afford school, so he dropped out of the University of Budapest in 1924. Yet, from 1924 to 1927, he published around 20 papers on mathematics and celestial mechanics. In correspondence about this research, Lichtenstein encouraged Wintner to continue his work toward a Ph.D. at the University of Leipzig, where Wintner received his degree in 1929 for work on "infinite matrices" on Hilbert Spaces, not on celestial mechanics. In his preface, Wintner claims that "this book could not have been written without the investigations of Levi-Civita and Birkhoff" where he acknowledges Birkhoff's help and encouragement. Wintner spent the 1929-1930 academic year in Rome with Levi-Civita, who we will meet on page 138. After his stay with Levi-Civita, in 1930 Wintner moved to Johns Hopkin where he remained for the rest of his professional life.

degenerate central configurations that are not part of a continuum of central configurations. He constructs one with equal masses on the vertices of an equilateral triangle, and a particle of mass m at the center. As m varies, it passes through a value causing the configuration to be degenerate.

2.3   A conjecture and a velocity decomposition

To verify the above comments about the role of central configurations in celestial mechanics, some technical material needs to be introduced. Technical material can be, well, boring. So to motivate these technical structures, which starts in Sect. 2.3.2, the material is described in terms of a specific unresolved conjecture about N-bodies that I advanced at a 1969 conference in Saõ Paulo, Brazil (Saari [74]). The conjecture involves the $I = \frac{1}{2} \sum_{j=1}^{N} m_j \mathbf{r}_j^2$ term (Eq. 2.15) that, as asserted by Cor. 2.1 (page 42), measures the diameter of the system.

2.3.1   Virial Theorem and the conjecture

The conjecture resulted from conversations with astronomers about a widely used tool in astronomy and astrophysics called the "Virial Theorem." For a discussion of this result, let me suggest the standard reference, which is Goldstein's classic book on mechanics [22]; for applications of this result to astrophysics, a reasonable source is Saslaw [98].

The Virial Theorem[8] states that if the positions and velocities are bounded for all time, that is if

$$I - O(1), \quad T = O(1) \text{ for } t \in [0, \infty) \tag{2.24}$$

where $O(1)$ means that the terms are bounded, then

$$\hat{T} \simeq \lim_{t \to \infty} \frac{1}{t} \int_0^t T(s)\, ds = \frac{1}{2} \lim_{t \to \infty} \frac{1}{t} \int_0^t U(s)\, ds \simeq \hat{U} \tag{2.25}$$

and $2\hat{T} = \hat{U}$

A glaring weakness of this theorem is that not only are the assumptions difficult to verify, but they probably are rarely, if ever, satisfied for actual astronomical and astrophysical problems of interest. Later (Sect. 4.5.2 starting on page 173, and, in particular, with Thm. 4.12) I show how the major difficulties can be removed with a significant generalization of this result by

---

[8] "Virial" is Latin for "force." This well-known result was found by Rudolph Claussius (1822-1888) as part of his study on the mechanical nature of heat during the nineteenth century development of thermodynamics.

using Tauberian Theorems. For instance, the assumptions on velocities and T turn out to be superfluous so they can be dismissed, and the conclusion holds even if the diameter of the system, I, expands, but not 'too fast."

While the Virial Theorem specifically refers to the time averages of the kinetic and self-potential, time averages are difficult to handle for practical applications to galaxies. As such, during the late 1960s (and even now), it was not uncommon to find people speculating that for certain galaxies the value of I "appears" to be nearly a constant: this made it reasonable to assume that I *is* a constant. The value added from this assumption is that it removes the time averages to assert that $2T = U$.

Lagrange Jacobi Equation

The stronger $2T = U$ conclusion is a direct consequence of the important *Lagrange-Jacobi equation*

$$I'' = U + 2h = T + h = 2T - U, \tag{2.26}$$

which describes how the self-potential of the system, U, affects the growth of the diameter of the system. To verify this equation, differentiate $I = \frac{1}{2}\sum m_j \mathbf{r}_j^2$ twice to obtain

$$I'' = \sum m_j \mathbf{v}_j^2 + \sum \mathbf{r}_j \cdot m_j \mathbf{r}_j'' = 2T + \sum \mathbf{r}_j \cdot \frac{\partial U}{\partial \mathbf{r}_j} = 2T - U$$

where the last expression uses Euler's Theorem. (For Euler's Theorem, see page 40.) The remaining terms in Eq. 2.26 come from using the energy integral (Eq. 2.6).

For our present purposes, notice how the assumption that I is a constant requires $I'' = 0$. In turn, according to Eq. 2.26, this means that $h < 0$ and $U = 2|h|$, or (from one of the other expressions) that $2T = U$. In other words, the assumption that I is a constant removes the time averages and leads to the desired and easier to use $2T = U$ conclusion.

Consequences of assumption

This "I equal to a constant" condition appears to be innocuous, even if it is not precisely correct. On the other hand, it also is reasonable to wonder whether the assumption carries unexpected, unintended consequences. Does anything restrictive or special happen for a constant moment of inertia? After all, the structure of the N-body problem and its analytic nature can be delicate. After exploring this problem, I conjectured:

Conjecture: If I equals a constant, then the motion of the N-body problem must be a relative equilibria—it is a coplanar central configuration that rotates with uniform circular motion.

If the conjecture is true, then the simplifying assumption of a constant I carries unwanted implications. No astronomers, for example, would want to assume anything that requires a coplanar galaxy with a rigid body rotation of the particles! In my presentation, I stated that the conjecture is true for the three-body problem, but that it was unsolved for the general N-body problem. Incidentally, while the conjecture is stated for the inverse-square force law, if true, it would hold for the inverse p-force law for $1 < p < 3$. It does not hold for $p = 3$.

Recently this conjecture received renewed attention when Z. Xia described it at a conference. Shortly after Xia's paper [115] entitled "Some of the problems Saari did not solve" appeared, I received requests for my three-body proof. As shown next, the proof for the three-body collinear case is trivial. But, to my embarrassment, I could not reconstruct the proof for the general three-body problem.

Collinear three body

To describe the trivial proof for the collinear three-body problem, use the fact proved in Prop. 2.1 (page 42) that

$$I = \frac{1}{2M} \sum_{j<k} m_j m_k r_{jk}^2$$

where $r_{jk} = |\mathbf{r}_j - \mathbf{r}_k|$ and M is the total mass. A second relationship comes from the Lagrange-Jacobi equation Eq. 2.26 requiring, as shown above, that $U = 2|h|$. A third relationship uses the fact that the three particles are on a line.

By differentiating the three equations

$$I = D, \quad U = 2|h|, \quad r_{1,2} + r_{2,3} - r_{1,3} = 0,$$

we obtain three independent equations that equal zero in the three variables $r'_{1,2}, r'_{2,3}, r'_{1,3}$. These equations force each of $r'_{1,2}, r'_{2,3}, r'_{1,3}$ to equal zero. Because the motion must be a relative equilibrium solution, it must keep the same configuration and same distances between particles for all time. Therefore, we have from the above (and from Thm. 2.5 on page 56) that the configuration is a central configuration.

This approach does not work for the collinear N-body problem for $N > 3$ because there are not enough equations. Thus a different argument involving the angular momentum is needed. The physical intuition is that if the configuration is not a central configuration, then some particles rotate faster than others to destroy the collinearity.[9]

In general?

What about the general three-body problem? Judged by the attention the conjecture has received by highly competent mathematicians,[10] this problem appears to be difficult to solve with standard methods. But while Llibre and Pina [39] and then Moeckel [50] found complicated computer assisted proofs for the three-body problem, at least as of now there are no analytic proofs. Part of the difficulty is the absence of tools that tell us how the system behaves with a constant I. (Equations and a discussion are given on page 80.) This search for an analytic proof helps motivate the more general N-body structures introduced below.

To continue the story, when my former students loaned me lecture notes from my 1969-70 course on the N-body problem to help write this book, I eagerly checked to find if this assertion appears. Great excitement: the theorem is in one set of notes! Great dismay: the proof reads, "Trivial." By examining the notes, it appears that I proved this result by showing for the three-body problem that if I is a constant, then $I \equiv c^2/4|h|$. Indeed, it turns out that $I \equiv c^2/4|h|$ is a necessary and sufficient condition for the conjecture to hold for any N.

Theorem 2.2 *For any $N \geq 3$, if I is a constant for $h < 0$, then*

$$\frac{c^2}{4|h|} \leq I.$$

If I is a constant, then a necessary and sufficient condition for the motion to be a coplanar rotating central configuration—a relative equilibria—is that

---

[9]After this was written, M. Santoprete told me that he, F. Diacu, and E. Perez used these arguments that I roughly outlined to them (while visiting the University of Victoria, BC, in September, 2002) to prove the conjecture for the collinear N-body problem. Their paper [17] will appear in the *Transactions* of the AMS. The argument that I briefly outlined for them, which differs from their approach, is described in Thm. 2.13 on page 68. Also see the discussion on page 79.

[10]A partial list includes Hernández-Garduno, Lawson, and Marsden [25], Llibre and Pina [39], McCord [43], Moeckel [50], Roberts [67], Diacu, Perez, Santoprete [17], and Santoprete [97].

$h < 0$ *and*

$$I \equiv \frac{c^2}{4|h|} \tag{2.27}$$

This proof (given on page 67) uses the system velocity decomposition that I refined in 1983 while visiting Hildeberto Cabral in Recife, Brazil (Saari [86, 87]): this decomposition, which is the real theme of this section, is described next.

2.3.2 The system velocity decomposition

We need to develop tools to better understand the above conjecture and, more importantly, the dynamics of the N-body problem. We need tools, for instance, to help verify the earlier comment that central configurations arise whenever the solution is homographic. All of this follows from the following velocity decomposition that plays an important role when we discuss collisions. Elsewhere (Saari [90]), this decomposition will be used to derive results about the evolution of N-body systems.

Motivation for what follows comes from introductory vector analysis where the velocity of a particle, $\mathbf{v}$, is decomposed into the component that changes the size of vector $\mathbf{r}$ and the component that changes the direction of $\mathbf{r}$. By use of scalar and cross products, the scalar and rotational components are, respectively,

$$\mathbf{v}_{scal} = r'\frac{\mathbf{r}}{r}, \quad \mathbf{v}_{rot} = \frac{\mathbf{r} \times \mathbf{v}}{r}. \tag{2.28}$$

System terms

The approach for N-body systems mimics what we do with $\mathbf{v}$. Toward this end, define the *system position and velocity* vectors to be, respectively,

$$\mathbf{R} = (\mathbf{r}_1, \mathbf{r}_2, \dots, \mathbf{r}_N), \quad \mathbf{V} = (\mathbf{v}_1, \mathbf{v}_2, \dots, \mathbf{v}_N). \tag{2.29}$$

To handle these $(\mathbb{R}^3)^N$ vectors, define the *system inner product* for $\mathbf{a} = (\mathbf{a}_1, \dots, \mathbf{a}_N), \mathbf{b} = (\mathbf{b}_1, \dots, \mathbf{b}_N) \in (\mathbb{R}^3)^N$ to be

$$< \mathbf{a}, \mathbf{b} >= \sum m_j \mathbf{a}_j \cdot \mathbf{b}_j. \tag{2.30}$$

With this notation, it follows immediately that

$$2I =< \mathbf{R}, \mathbf{R} >, \quad 2T =< \mathbf{V}, \mathbf{V} > . \tag{2.31}$$

As the gradient is defined relative to the inner product, the *system gradient* ∇_s converts the earlier equations of motion Eq. 2.5 into the form

$$\mathbf{R}'' = \nabla_s U = (\frac{1}{m_1}\frac{\partial U}{\partial \mathbf{r}_1}, \ldots, \frac{1}{m_N}\frac{\partial U}{\partial \mathbf{r}_N}).$$

By using these terms, the Eq. 2.8 (page 35) definition of a central configuration is that there is a scalar λ so that

$$\lambda \mathbf{R} = \nabla_s U. \tag{2.32}$$

System velocity components

To mimick the Eq. 2.28 decomposition, the easiest $\mathbf{V}$ component to find is the one that just changes the size of the configuration $\mathbf{R}$. As in Eq. 2.28, this *system scalar velocity* is $\mathbf{W}_{scal} = \lambda\frac{\mathbf{R}}{R}$ where

$$\lambda = <\mathbf{V}, \frac{\mathbf{R}}{R}> = R'.$$

Thus,

$$\mathbf{W}_{scal} = R'\frac{\mathbf{R}}{R}. \tag{2.33}$$

To find the $\mathbf{V}$ component corresponding to the rotational term in Eq. 2.28, notice that a way to find $\mathbf{v}_{rot}$ in Eq. 2.28 is to first rotate $\mathbf{r}$ in all possible ways to define a sphere. Next find the tangent plane to this sphere at $\mathbf{r}$: the velocity component $\mathbf{v}_{rot}$ is the projection of $\mathbf{v}$ to this tangent plane. To carry out a similar argument for the system vector $\mathbf{R}$, let

$$\mathcal{M}_{\mathbf{R}} = \{\Omega(\mathbf{R}) = (\Omega(\mathbf{r}_1), \ldots, \Omega(\mathbf{r}_N)) \,|\, \Omega \in SO(3)\}$$

be the manifold that consists of all rigid body rotations of $\mathbf{R}$. The component of $\mathbf{V}$ in the tangent space $T_{\mathbf{R}}\mathcal{M}_{\mathbf{R}}$, then, is the system velocity component that rotates configuration $\mathbf{R}$ as a rigid body. To find this system rotational velocity $\mathbf{W}_{rot}$, we need to find a basis for $T_{\mathbf{R}}\mathcal{M}_{\mathbf{R}}$.

A basis for $T_{\mathbf{R}}\mathcal{M}_{\mathbf{R}}$ involves specifying three independent axes of rotation for the configuration. Letting $\mathbf{E}_j = (\mathbf{e}_j, \ldots, \mathbf{e}_j) \in (\mathbb{R}^3)^N$, where $\mathbf{e}_j \in \mathbb{R}^3$ is the unit vector with unity in the jth component, a basis is

$$\{\mathbf{E}_j \times \mathbf{R} := (\mathbf{e}_j \times \mathbf{r}_1, \ldots, \mathbf{e}_j \times \mathbf{r}_N)\}_{j=1}^3. \tag{2.34}$$

Thus the system rotational velocity is

$$\mathbf{W}_{rot} = \sum_{j=1}^3 s_j(\mathbf{E}_j \times \mathbf{R}) = \mathbf{S} \times \mathbf{R} \tag{2.35}$$

where $\mathbf{S} = (\mathbf{s}, \ldots, \mathbf{s})$ and $\mathbf{s} = (s_1, s_2, s_3) \in \mathbb{R}^3$ is the instantaneous axis of rotation for the configuration.

As an aside, another use of the vectors $\mathbf{E}_j$ is to represent the center of mass integrals (Eq. 2.2, page 33) in the compact form of

$$< \mathbf{E}_j, \mathbf{R} > = < \mathbf{E}_j, \mathbf{V} > = 0 \text{ for } j = 1, 2, 3. \tag{2.36}$$

Returning to the system rotational velocity, a straightforward computation shows that

$$< \mathbf{V}, \frac{\mathbf{E}_j \times \mathbf{R}}{||\mathbf{E}_j \times \mathbf{R}||} > = \frac{< \mathbf{E}_j, \mathbf{R} \times \mathbf{V} >}{||\mathbf{E}_j \times \mathbf{R}||} = \frac{\mathbf{e}_j \cdot \mathbf{c}}{||\mathbf{E}_j \times \mathbf{R}||} \tag{2.37}$$

where $\mathbf{c} - (0, 0, c)$ is the constant of angular momentum from Eq. 2.3 (page 33). A problem for non-coplanar motion is that $\{\mathbf{E}_j \times \mathbf{R}\}_{j=1}^3$ is not an orthogonal basis, and problems arise when computing the denominator term $||\mathbf{E}_j \times \mathbf{R}||$. But everything is simple with coplanar solutions because the decomposition involves only the $\mathbf{E}_3 \times \mathbf{R}$ term where the configuration rotates 90°: here $||\mathbf{E}_3 \times \mathbf{R}|| = R$.

To review, the $\mathbf{W}_{scal}$ velocity component changes the scale of the $\mathbf{R}$ configuration and the $\mathbf{W}_{rot}$ component changes the orientation. What remains, the *configurational velocity* $\mathbf{W}_{config} = \mathbf{V} - (\mathbf{W}_{scal} + \mathbf{W}_{rot})$, is the $\mathbf{V}$ component that changes the shape of the N-body configuration. A characterization of $\mathbf{W}_{rot}$ and $\mathbf{W}_{scal}$ is given next; a representation for $\mathbf{W}_{config}$ is given at the end of this chapter. In what follows let $\mathbf{X}, \mathbf{Y}$, and $\mathbf{Z}$ be the obvious components of $\mathbf{R}$. To simplify the notation for one statement in the next theorem, let $\sigma_X(\mathbf{Y})$ and $\sigma_X(\mathbf{Z})$ be where the components are shifted to the X position; e.g., $\sigma_X(\mathbf{Z}) = ((z_1, 0, 0), \ldots, (z_N, 0, 0))$.

Theorem 2.3 *The N-body system velocity can be expressed as*

$$\mathbf{V} = \mathbf{W}_{rot} + \mathbf{W}_{scal} + \mathbf{W}_{config}. \tag{2.38}$$

where

$$\mathbf{W}_{scal} = R' \frac{\mathbf{R}}{R}, \quad \mathbf{W}_{config} = \mathbf{V} - (\mathbf{W}_{rot} + \mathbf{W}_{scal}). \tag{2.39}$$

For the coplanar N-body problem

$$\mathbf{W}_{rot} = \frac{c(\mathbf{E}_3 \times \mathbf{R})}{R^2} \text{ and } \mathbf{s} = \frac{c\mathbf{e}_3}{R^2}. \tag{2.40}$$

For the non-coplanar problem, $\mathbf{W}_{rot}$ is defined by $\mathcal{A}\mathbf{s} = \mathbf{c}$ where

$$\mathcal{A} = \begin{pmatrix} R^2 - X^2 & - < \mathbf{X}, \sigma_X(\mathbf{Y}) > & - < \mathbf{X}, \sigma_X(\mathbf{Z}) > \\ - < \sigma_X(\mathbf{Y}), \mathbf{X} > & R^2 - Y^2 & - < \sigma_X(\mathbf{Y}), \sigma_X(\mathbf{Z}) > \\ - < \sigma_X(\mathbf{Z}), \mathbf{X} > & - < \sigma_X(\mathbf{Z}), \sigma_X(\mathbf{Y}) > & R^2 - Z^2 \end{pmatrix} \tag{2.41}$$

In particular,

$$\mathbf{W}_{rot}^2 = (\mathbf{S} \times \mathbf{R})^2 \geq \frac{c^2}{R^2 - Z^2} = \frac{c^2}{X^2 + Y^2}. \qquad (2.42)$$

The $\mathbf{W}_j$ terms are pairwise orthogonal (with respect to $< -, - >$), so the conservation of energy integral (Eq. 2.6) can be expressed as

$$\mathbf{V}^2 = \mathbf{W}_{rot}^2 + \mathbf{W}_{scal}^2 + \mathbf{W}_{config}^2 = 2(U + h). \qquad (2.43)$$

As we will see, Eq. 2.43 plays an important role in analyzing N-body motion. It should as it is the energy integral described in terms of the velocity decomposition.

Proof: That the $\mathbf{W}_j$ vectors are orthogonal follows from the fact that $\mathbf{R}$ is orthogonal to $T_{\mathbf{R}}\mathcal{M}_{\mathbf{R}}$ and the definition of $\mathbf{W}_{config}$. The proof of Eq. 2.40 follows immediately from Eqs. 2.35, 2.37.

What complicates the description of the rotational velocity $\mathbf{W}_{rot}$ is that the basis in Eq. 2.34 is not orthogonal. But $\|\mathbf{E}_3 \times \mathbf{R}\|^2 = R^2 - Z^2$, so it follows from Eq. 2.37 that

$$< \mathbf{W}_{rot}, \frac{\mathbf{E}_j \times \mathbf{R}}{\|\mathbf{E}_j \times \mathbf{R}\|} > = \begin{cases} 0 & \text{if } j = 1, 2 \\ \frac{c}{\|\mathbf{E}_3 \times \mathbf{R}\|} = \frac{c}{\sqrt{R^2 - Z^2}} & \text{if } j = 3 \end{cases} \qquad (2.44)$$

Using Eq. 2.44 with the three equations $< \mathbf{W}_{rot}, \mathbf{E}_j \times \mathbf{R} >$, $j = 1, 2, 3$, that come from Eq. 2.35 leads to the matrix in Eq. 2.41.

Finally, according to Eq. 2.44, the only component of $\mathbf{W}_{rot}$ in the $\mathbf{E}_3 \times \mathbf{R}$ direction is $[c\mathbf{E}_3 \times \mathbf{R}]/[R^2 - Z^2]$. Therefore, we obtain Eq. 2.42:

$$\mathbf{W}_{rot}^2 \geq [\frac{c\mathbf{E}_3 \times \mathbf{R}}{R^2 - Z^2}]^2 = \frac{c^2}{R^2 - Z^2}.$$

From Eq. 2.31 and Eq. 2.6, we have that $V^2 = 2T = 2(U + h)$. Using Eq. 2.38 and the orthogonality of the $\mathbf{W}_j$ terms, Eq. 2.43 follows. $\square$

2.3.3 Central configurations and the velocity decomposition

What does all of this have to do with central configurations? To explain, Thm. 2.3 describes how the system velocity $\mathbf{V}$ changes the N-body configuration $\mathbf{R}$ where, for instance, $\mathbf{W}_{rot}$ rotates the configuration as a rigid body about an instantaneous axis $\mathbf{s}$ while keeping the configuration $\mathbf{R}$ unchanged. As the two velocity components $\mathbf{W}_{scal}$ and $\mathbf{W}_{rot}$ change the scale and orientation of the configuration, they keep the configuration in the same Euler similarity class: if they are the only $\mathbf{V}$ components, the solution is homographic. The $\mathbf{W}_{config}$ term, however, changes the shape of the configuration; i.e., this $\mathbf{V}$ component forces $\mathbf{R}$ to change Euler similarity classes.

Recall those earlier comments suggesting that we must anticipate central configurations whenever the configuration remains unchanged: these assertions reduce to understanding what happens if $\mathbf{W}_{config} = \mathbf{0}$. Indeed, the Sect. 2.2.1 assertions about central configurations involve showing either that $\mathbf{W}_{config} \equiv \mathbf{0}$, or that $\mathbf{W}_{config}, \mathbf{W}'_{config} \to \mathbf{0}$. So a way to prove my conjecture that a constant value of I requires the motion to be a relative equilibrium solution is to prove that $\mathbf{W}_{scal} \equiv \mathbf{0}$ (which is immediate as $R' \equiv 0$) and that $\mathbf{W}_{config} \equiv \mathbf{0}$.

Before moving on, let me offer an useful result connecting $\mathbf{W}_{config}$ with the configurational measure (see page 42) RU, or $\sqrt{I}U$.

Theorem 2.4 *For the coplanar N-body problem,*

$$[(R\mathbf{W}_{config})^2]' = 2R(RU)'. \tag{2.45}$$

The theorem states what we should expect: changes in the configuration that alter the configurational measure's RU value must be associated with changes in the $(R\mathbf{W}_{config})^2$ value.

Proof: Multiplying both sides of Eq. 2.43 by R, using the $\mathbf{W}_j$ values from Thm. 2.3, and solving for RU leads to

$$\frac{c^2}{R} + R(R')^2 + R\mathbf{W}^2_{config} \quad 2hR = 2RU.$$

By differentiating, we obtain

$$-\frac{c^2 R'}{R^2} + R'(R')^2 + R2R'R'' + R'\mathbf{W}^2_{config} + R(\mathbf{W}^2_{config})' - 2hR' = 2(RU)'$$

or

$$R'(\ [\tfrac{c^2}{R^2} + (R')^2 + \mathbf{W}^2_{config}] + \ 2\{RR'' + (R')^2\} - 2h + 2\mathbf{W}^2_{config}]$$
$$+R(\mathbf{W}^2_{config})' = 2(RU)'.$$

According to Eq. 2.43, the term in the "[ ]" brackets equals V^2; the term in the "{ }" brackets is I'' so, according to Eq. 2.26, it equals $T + h$: twice this value is $V^2 + 2h$. After some obvious cancellations, we obtain

$$2R'\mathbf{W}^2_{config} + R(\mathbf{W}^2_{config})' = 2(RU)'.$$

The conclusion follows by multiplying this expression by R and recognizing that $2RR'[\mathbf{W}^2_{config} + R^2[\mathbf{W}^2_{config}]' = [R^2\mathbf{W}^2_{config}]'$. $\square$

2.3.4   Motion preserving an Euler similarity class

To support the comment that central configurations should be anticipated whenever $\mathbf{W}_{config} \equiv \mathbf{0}$, recall that $\mathbf{W}_{config} \equiv \mathbf{0}$ forces the particles to retain the same configuration so the motion remains in the same Euler similarity class. As already advertised, the next classical result (expressed in terms of the $\mathbf{W}_j$'s) states that the configuration must be a central configuration.

Theorem 2.5 *If $\mathbf{W}_{config} \equiv \mathbf{0}$ for a solution of the coplanar N-body problem, the particles form a central configuration. That is, for the coplanar problem, a homographic solution is defined by a central configuration.*

While three-body versions of the results in this section have been known since Lagrange, and certain N-body versions since the work of Pizzetti[11] [59], Winter explains that one rarely sees the proofs because they are complicated and messy. The following simpler proofs underscore the power of the velocity decomposition: the arguments essentially reduce to calculus.

Proof: If $\mathbf{W}_{config} \equiv \mathbf{0}$, then $\mathbf{V} = \mathbf{W}_{rot} + \mathbf{W}_{scal}$, so $\mathbf{V}' = \mathbf{R}'' = \nabla_s U$ becomes

$$\mathbf{R}'' = \mathbf{W}'_{rot} + \mathbf{W}'_{scal} = \nabla_s U. \qquad (2.46)$$

Differentiating $\mathbf{W}_{rot} = \frac{c}{R^2}\mathbf{E}_3 \times \mathbf{R}$ and using the decomposition of $\mathbf{V}$ leads to $\mathbf{W}'_{rot} = -2\frac{cR'}{R^3}\mathbf{E}_3 \times \mathbf{R} + \frac{c}{R^2}\mathbf{E}_3 \times (\frac{c}{R^2}\mathbf{E}_3 \times \mathbf{R} + \frac{R'}{R}\mathbf{R})$. As $\mathbf{E}_3 \times (\mathbf{E}_3 \times \mathbf{R}) = -\mathbf{R}$,

$$\mathbf{W}'_{rot} = -\frac{c^2}{R^4}\mathbf{R} - \frac{cR'}{R^3}\mathbf{E}_3 \times \mathbf{R}. \qquad (2.47)$$

Similarly, as $\mathbf{W}_{scal} = \frac{R'}{R}\mathbf{R}$, we have that

$$\mathbf{W}'_{scal} = \frac{(R''R - R'^2)}{R^2}\mathbf{R} + \frac{R'}{R}[\frac{c}{R^2}\mathbf{E}_3 \times \mathbf{R} + \frac{R'}{R}\mathbf{R}] = \frac{R''}{R}\mathbf{R} + \frac{cR'}{R^3}\mathbf{E}_3 \times \mathbf{R}. \qquad (2.48)$$

By substituting Eqs. 2.47, 2.48 into Eq. 2.46 we obtain the desired conclusion (see Eq. 2.32) that

$$\mathbf{R}'' = [\frac{R''}{R} - \frac{c^2}{R^4}]\mathbf{R} = \nabla_s U. \qquad \square$$

---

[11]P. Pizzetti (1860-1918) made several contributions to celestial mechanics and mathematical astronomy including the Somigliana-Pizzetti model in geodesy that is used in terrestrial investigations. (Pizzetti developed the theory of the equipotential ellipsoid of revolution in 1894, and Somigliana added to it in 1927.) An honor accorded Pizzetti is a crater on the moon that is named after him.

The $\mathbf{R}'' = \nabla_s U = -\frac{A}{R^3}\mathbf{R}$ equation suggests that for each particle, the N-body dynamics within this similarity class is precisely that of the two-body problem. To suggest why, notice that differentiating $R^2 = \mathbf{R}^2$ twice leads to $R'' = \frac{\mathbf{R}}{R} \cdot \mathbf{R}'' + \frac{[\mathbf{V}^2 - (R')^2]}{R}$ or, because $\mathbf{V}^2 - (R')^2 = \mathbf{W}_{rot}^2$ (as $\mathbf{W}_{config} \equiv \mathbf{0}$),

$$R'' = \frac{\mathbf{R}}{R} \cdot \nabla_s U + \frac{c^2}{R^3} = -\frac{A}{R^2} + \frac{c^2}{R^3}. \tag{2.49}$$

But Eq. 2.49 is precisely the equation of motion for $r = \|\mathbf{r}_2 - \mathbf{r}_1\|$ in the two body problem, so we must anticipate a connection. To find this connection, start with a central configuration $\mathbf{R}$ and assign initial conditions for $\mathbf{V}$ where $\mathbf{W}_{config} = \mathbf{0}$ (i.e., the velocity initial conditions are given by $\mathbf{W}_{scal} + \mathbf{W}_{rot}$).[12] It turns out that each particle follows an ellipse, or a parabola, or a hyperbola, or is rectilinear.

Theorem 2.6 *Assume that the configuration of an N-body homographic solution always is a central configuration. For a three-dimensional configuration, $\mathbf{c} = \mathbf{0}$ and the motion is homothetic. If $\mathbf{c} \neq \mathbf{0}$, then the motion is coplanar in the invariable plane. In any case, each particle moves as though its motion is governed by the two-body problem.*

Proof: According to Thm. 2.1, if the N-body configuration always is a central configuration, then $\mathbf{R}'' = -\frac{U}{2I}\mathbf{R} = -\frac{(RU)}{R^3}\mathbf{R}$. As the configurational measure RU is homogenous of degree 0, it is a constant, say A, that is determined by the fixed configuration. That is, $\mathbf{R}'' = -\frac{A\mathbf{R}}{R^3}$.

With a fixed configuration, each r_j must be a fixed multiple of the value of R; i.e., each $r_j = B_j R$ for some constant B_j. Thus, for each $j = 1, \ldots, N$, the equations of motion are

$$\mathbf{r}_j'' = -\frac{A_j' \mathbf{r}_j}{r_j^3}$$

where $A' = \frac{A}{B_j^3}$ is a positive constant. As this is the equation of the two-body problem, the last assertion is proved.

The configuration remains fixed for all time, so $\mathbf{W}_{config} \equiv \mathbf{0}$. Thus, at any instant of time, the initial conditions are given by $\mathbf{W}_{rot}$ and $\mathbf{W}_{scal}$. As $\mathbf{W}_{rot} = (\mathbf{s} \times \mathbf{r}_1, \ldots, \mathbf{s} \times \mathbf{r}_N)$ for some $\mathbf{s} \in \mathbb{R}^3$, it follows from the angular

---

[12]Thus, the velocity components for each body are proportional to the r_j distances. This is because $\mathbf{W}_{scal} = (\frac{R'}{R}\mathbf{r}_1, \ldots, \frac{R'}{R}\mathbf{r}_N)$ while $\mathbf{W}_{rot} = \frac{c}{R^2}(\mathbf{e}_3 \times \mathbf{r}_1, \ldots, \mathbf{e}_3 \times \mathbf{r}_N)$.

momentum integral and the constant $\mathbf{c} = (0, 0, c)$ that $\mathbf{s}/s = \mathbf{e}_3$. Thus, if the configuration is one or two-dimensional, it is in the invariable plane.

It remains to show that if $\mathbf{R}$ defines a three-dimensional configuration, then $\mathbf{c} = \mathbf{0}$ and, because the system is absent any rigid body motion, the motion is homothetic. The intuition behind the argument is natural. As the configuration rotates about the z axis, we should expect the force law to drag $\mathbf{z}$ components to the $x - y$ invariable plane: this change in the configuration creates a contradiction. To prove the assertion in this manner, observe that if the motion preserves a central configuration and if $\mathbf{Z} \neq \mathbf{0}$, then $Z = AR$ for some positive constant A: this means that $\frac{R''}{R} \equiv \frac{Z''}{Z}$. By computing the $\frac{R''}{R} \equiv \frac{Z''}{Z}$ terms, it will follow that $\mathbf{c} = \mathbf{0}$.

Using the approach leading to Eq. 2.49, we have

$$R'' = \frac{\mathbf{R}}{R} \cdot \mathbf{R}'' + \frac{c^2}{R(R^2 - Z^2)} = -\frac{(RU)}{R^2} + \frac{c^2}{R(R^2 - Z^2)} \tag{2.50}$$

where the last equality uses $\mathbf{R} \cdot \nabla_s U = -U$. As the configuration is preserved and $Z = AR$, it follows from the homogeneity of degree zero of RU that there are two positive constants A_1, A_2 so that Eq. 2.50 can be expressed as

$$R'' = -\frac{A_1}{R^2} + \frac{c^2 A_2}{R^3}. \tag{2.51}$$

Combining $\mathbf{R}'' = \lambda \mathbf{R} = \nabla_s U$ with $\frac{R''}{R} = \frac{Z''}{Z}$, we have that $\mathbf{Z}'' = \frac{R''}{R} \mathbf{Z}$, and $\mathbf{Z}''$ is given by the z components of $\nabla_s U$. (The particle with the largest $z > 0$ value must have a negative z-component in $\nabla_s U$, so $R'' < 0$.) The $Z^2 = \mathbf{Z}^2$ equality leads to $Z'' = \frac{\mathbf{Z}}{Z} \cdot \mathbf{Z}'' + \frac{[(\mathbf{Z}')^2 - (Z')^2]}{Z}$. But as the configuration rotates about the z-axis and is preserved, there is a constant $\mathbf{a} = ((0, 0, a_1), \dots, (0, 0, a_N))$ so that $\mathbf{Z}(t) = \lambda(t)\mathbf{a}$. Thus, $\mathbf{Z}' = \lambda'\mathbf{a}$ and $Z' = \lambda' a$, so $Z'' = \frac{\mathbf{Z}}{Z} \cdot \mathbf{Z}'' = \frac{\mathbf{Z}}{Z} \cdot \nabla_s U$, which, because the configuration is preserved, can be expressed as

$$Z'' = -\frac{B}{Z^2} \text{ for some constant } B > 0. \tag{2.52}$$

Comparing Eqs. 2.51, 2.52 leads to the desired conclusion that either $\mathbf{Z} \equiv \mathbf{0}$, or if $\frac{R''}{R} \equiv \frac{Z''}{Z}$, then $\mathbf{c} = \mathbf{0}$. $\square$

The dynamics

The simplicity of the proof of Thm. 2.5 hides the impact of the $\mathbf{W}_{config} \equiv \mathbf{0}$ assumption on the dynamics. To explain, first notice that the dimension of the space for the configurational velocity is $3N - 7$ for the general N-body problem, and $2N - 4$ for the coplanar problem. These values come from the dimensions of the two choices of $\mathbf{V}$ spaces minus the dimensions for the center of mass, rotational velocity, and radial $\mathbf{R}$ direction. Let $\{\mathbf{U}_j\}_{j=1}$ be a basis for this space.[13] Because $\mathbf{W}_{config} = \sum_j \beta_j \mathbf{U}_j$, we have that

$$\mathbf{W}'_{config} = \sum \beta'_j \mathbf{U}_j + \sum_j \beta \mathbf{U}'_j. \qquad (2.53)$$

Thus, if either $< \mathbf{W}'_{rot}, \mathbf{U}_j > \neq 0$, or $< \mathbf{W}'_{scal}, \mathbf{U}_j > \neq 0$, then $\mathbf{W}'_{config} \neq \mathbf{0}$. In other words, if $\mathbf{W}'_{rot}$ cannot be expressed as

$$\mathbf{W}'_{rot} = \alpha \mathbf{R} + \mathbf{A} \times \mathbf{R}, \qquad (2.54)$$

for scalar α and $\mathbf{A} \in \mathbb{R}^3$, then the extra $\mathbf{W}'_{rot}$ components are in the configurational velocity space: they contribute to the dynamics of $\mathbf{W}_{config}$ and ensure that the shape of the configuration must change.

The $\mathbf{W}_{config} \equiv \mathbf{0}$ assumption, then, restricts the $\mathbf{W}_{rot}$ behavior as specified in Thm. 2.5. Another way to make this point is to note that the general coplanar computation replaces Eq. 2.47 with

$$\mathbf{W}'_{rot} = -\frac{c^2}{R^4}\mathbf{R} - \frac{cR'}{R^3}\mathbf{F}_3 \times \mathbf{R} + \frac{c}{R^2}\mathbf{F}_3 \times \mathbf{W}_{config} : \qquad (2.55)$$

portions of the last term contribute to changes in the shape of the configuration, and, conversely, $\mathbf{W}_{config}$ contributes to changes in the rotational velocity. The same effect occurs, but not as dramatically for $\mathbf{W}_{scal}$.

The symbiotic exchange among the $\mathbf{W}_j$'s can be captured by using the orthogonality conditions such as $< \mathbf{W}_{rot}, \mathbf{W}_{config} >= 0$. By differentiating, we have that $< \mathbf{W}'_{rot}, \mathbf{W}_{config} > + < \mathbf{W}_{rot}, \mathbf{W}'_{config} >= 0$. In other words, the contribution $\mathbf{W}'_{rot}$ makes to $\mathbf{W}_{config}$—toward changing the shape of the configuration—is reciprocated: $\mathbf{W}'_{config}$ affects $\mathbf{W}_{rot}$ and the rotational behavior of the system. This makes sense; e.g., if the configuration becomes "tighter," we expect a faster spin. To sample other relationships, because $< \mathbf{R}, \mathbf{W}_{rot} >=< \mathbf{R}, \mathbf{W}_{scal} >= 0$, we have

$$\mathbf{W}^2_{rot} = - < \mathbf{R}, \mathbf{W}'_{rot} >, \quad \mathbf{W}^2_{scal} = - < \mathbf{R}, \mathbf{W}'_{scal} > . \qquad (2.56)$$

---

[13] An orthogonal basis for the coplanar N-body problem is given in Sect. 2.5 (page 69).

Representation for $\nabla_s U$

The configurational velocity space creates a new way (that begs to be exploited) to represent $\nabla_s U$. To see the connection, observe that the dimension of the space in which $\nabla_s U$ resides is one more than that for the configurational velocity. More striking are the similarities: both exclude rotations and the center of mass. Thus, by making standard identifications between velocity and position basis vectors, a basis for $\nabla_s U$ is given by $\{\mathbf{R}, \{\mathbf{U}_j\}_j\}$. By use of Euler's Theorem, there exist scalar functions so that

$$\nabla_s U = -\frac{U}{R^2}\mathbf{R} + \sum_j \alpha_j \mathbf{U}_j. \qquad (2.57)$$

This surprising representation states that the lead term for $\nabla_s U$ *always* is the $-\frac{U}{R^2}\mathbf{R} = \nabla_s U$ term from central configurations. In other words, a central configuration occurs iff $\nabla_s U$ has no components orthogonal to $\mathbf{R}$. It follows from Eq. 2.57, for instance, that

> *the N-body configuration is a central configuration iff there are no $\mathbf{U}_j$ terms on the $\mathbf{R}''$ side of the equations of motion.*

To have a central configuration, then, the $\mathbf{U}_j$ terms on the $\mathbf{R}''$ side of the equation must cancel. But non-zero $\mathbf{U}_j$ terms in $\mathbf{R}''$ (or in $\nabla_s U$) will unleash $\mathbf{W}_{config}$ to force configurational changes. To prove that this cancellation can occur, let $\mathbf{R}$ be a central configuration, and assign initial conditions where $\mathbf{W}_{config}$ and/or $\mathbf{W}'_{config}$ do not equal to zero. At the initial time, $t = 0$, the $\mathbf{U}_j$ terms on the $\mathbf{R}''$ side of the equation must cancel to permit the central configuration, but these $\mathbf{U}_j$ components will unleash $\mathbf{W}_{config}$ to ensure subsequent configurational changes.

Non-coplanar setting

This dynamical exchange among the $\mathbf{W}_j$ terms occurs in the proof of the next result that extends Thm. 2.5 to the non-coplanar case.

Theorem 2.7 *For the general N-body problem, if $\mathbf{W}_{config} \equiv \mathbf{0}$ (so the motion stays in the same Euler similarity class), then the particles always define a central configuration. If $\mathbf{c} \neq \mathbf{0}$, the motion is coplanar. If the configuration is not coplanar, then $\mathbf{c} = \mathbf{0}$ and the motion is homothetic.*

Proof: By differentiating $\mathbf{W}_{rot} = \mathbf{S} \times \mathbf{R}$ and collecting terms, we have

$$\mathbf{W}'_{rot} = [\mathbf{S}' + \frac{R'}{R}\mathbf{S}] \times \mathbf{R} + \mathbf{S} \times (\mathbf{S} \times \mathbf{R})$$

while

$$\mathbf{W}'_{scal} = \frac{R''}{R}\mathbf{R} + \frac{R'}{R}\mathbf{S} \times \mathbf{R}.$$

As $\mathbf{W}_{config} \equiv \mathbf{0}$, $\mathbf{V} = \mathbf{W}_{rot} + \mathbf{W}_{scal}$, so

$$\mathbf{R}'' = [\mathbf{S}' + 2\frac{R'}{R}\mathbf{S}] \times \mathbf{R} + \mathbf{S} \times (\mathbf{S} \times \mathbf{R}) + \frac{R''}{R}\mathbf{R} = \nabla_s U. \qquad (2.58)$$

As U is invariant with respect to rotations, $\nabla_s U$ is orthogonal to $\mathbf{S}^* \times \mathbf{R}$ for any choice of $\mathbf{S}^* = (\mathbf{s}^*, \ldots, \mathbf{s}^*)$. Consequently, all rotations terms in the middle portion of Eq. 2.58 must cancel. The questionable term is $\mathbf{S} \times (\mathbf{S} \times \mathbf{R}) = -\alpha(\mathbf{R} - \mathbf{Z}_s)$ where $\mathbf{Z}_s$ consists of the $\mathbf{R}$ components in the s/s direction and α is a positive constant. If $\mathbf{Z}_s = \mathbf{0}$, then $[\mathbf{S}' + 2\frac{R''}{R}\mathbf{S}] \times \mathbf{R} \equiv \mathbf{0}$ and

$$\mathbf{R}'' = [-\alpha + \frac{R''}{R}]\mathbf{R}; \qquad (2.59)$$

i.e., the motion preserves a central configuration. The conclusions now follows from Thm. 2.6.

If $\mathbf{Z}_s \neq \mathbf{0}$, then it must be that $\mathbf{S} \times (\mathbf{S} \times \mathbf{R}) = \alpha(\mathbf{R} - \mathbf{Z}_s) = \beta\mathbf{R} + (\mathbf{A} \times \mathbf{R})$ for some $\mathbf{A} = (\mathbf{a}, \ldots, \mathbf{a})$; i.e., $\mathbf{S} \times (\mathbf{S} \times \mathbf{R})$ can have no terms in the configurational velocity direction. (There are no other terms in Eq. 2.58 to cancel any $\mathbf{S} \times (\mathbf{S} \times \mathbf{R})$ terms in this direction, so a configurational velocity component would make $\mathbf{W}_{config} \neq \mathbf{0}$.) In any case, as the rotational terms must cancel, the equations of motion become that of Eq. 2.59, and the conclusion follows from Thm. 2.6. $\square$

2.3.5 Sundman inequality

Another valued application of the velocity decomposition is that it extends an inequality discovered by Sundman to an equality. The Sundman inequality plays a key role in the N-body problem as manifested by the fact it keeps arising in celestial mechanics including the Chap. 4 discussion about the behavior of collisions. I introduce it now to show how this inequality is a crude approximation of the Eq. 2.43 velocity decomposition. Indeed, this connection with the energy integral explains why Sundman's discovery has played such a central role in the N-body problem.

Theorem 2.8 *(Sundman) All solutions of the N-body problem satisfy*

$$\mathbf{c}^2 + I'^2 \leq 4IT = 4I(U + h). \qquad (2.60)$$

To suggest how this inequality provides information about the dynamics and configurations, use the two extremes of Eq. 2.60 to solve for the *configurational measure* in the $\sqrt{I}U$ form. to obtain

$$G(\sqrt{I}, I') = \frac{c^2 + I'^2}{4\sqrt{I}} - h\sqrt{I} \leq \sqrt{I}U. \tag{2.61}$$

According to Eq. 2.61, the measure of the configuration formed by the particles is bounded below by the angular momentum c^2 and the diameter of the system $\sqrt{I}$. In particular, by finding the minimum value for $G(x, 0)$ for $h < 0$, which occurs at $I = \frac{c^2}{4}|h|$ (recall that we first encountered this I value in Thm. 2.2), we have that (Saari [82])

$$\text{for } h < 0, \text{ it always is true that } c|h| \leq \sqrt{I}U. \tag{2.62}$$

These connections are explored later (e.g., page 157 and [90]): our current goal is to prove Sundman's inequality. A stronger version of Sundman's inequality[14] follows. To prove Eq. 2.60, add the two inequalities of Eq. 2.63.

Theorem 2.9 *All solutions of the N-body problem satisfy the inequalities*

$$(I')^2 \leq 4IQ, \quad c^2 \leq 4I(T - Q), \tag{2.63}$$

where $Q = \frac{1}{2} \sum m_j (r'_j)^2$.

Proof: As

$$I' = \sum m_j r_j r'_j = \sum [\sqrt{m_j} r_j][\sqrt{m_j} r'_j],$$

the first inequality follows immediately from the Cauchy-Schwartz inequality. To derive the second inequality, because

$$|\mathbf{c}| \leq \sum m_j r_j \frac{|\mathbf{r}_j \times \mathbf{v}_j|}{r_j} = \sum [\sqrt{m_j} r_j][\sqrt{m_j} \frac{|\mathbf{r}_j \times \mathbf{v}_j|}{r_j}],$$

we have from the Cauchy-Schwartz inequality that

$$c^2 \leq 2I \sum m_j \frac{(\mathbf{r}_j \times \mathbf{v}_j)^2}{r_j^2}.$$

But $r^2 v_j^2 = (r_j r'_j)^2 + (\mathbf{r}_j \times \mathbf{v}_j)^2$, so each term in the summation can be expressed as $m_j(v_j^2 - (r'_j)^2)$, and the conclusion follows. $\square$

---

[14]Hulkower's notes [27] suggest that I developed it around 1969.

There is a slight problem should some $r_j = 0$. But as solutions are analytic, either $r_j \equiv 0$, which causes no problem as the body is excluded from the summations, or the isolated zeros are handled in a normal manner. An alternative approach is to use $Q = \frac{1}{2M} \sum_{j<k} m_j m_k (r'_{j,k})^2$. Indeed, I used this alternative approach in analyzing my earlier conjecture.

To indicate how Thm. 2.9 relates to my conjecture (page 47) that a constant I requires a relative equilibrium motion, notice from the energy integral $T = U + h$ and the $U \equiv -2h$ condition, which accompanies a constant I, that $T \equiv -h$. If $I \equiv \frac{c^2}{2|h|}$, we have from Eq. 2.63 that

$$\mathbf{c}^2 \leq 4IT - 4IQ = 4(\frac{c^2}{4|h|})(|h|) - 4IQ,$$

or $4IQ \leq 0$. As the last inequality requires $Q \equiv 0$, or each $r'_{j,k} \equiv 0$, the configuration must remain unchanged. The fixed configuration cannot be homothetic, so the conclusion would follow from Thm. 2.7. As stated earlier, about the time I made the conjecture, I recall using the specific structure of the three-body problem to show that $I \equiv \frac{c^2}{4|h|}$. But, I cannot recover this proof, and I am not sure whether the argument obtaining this I value extends to the N-body problem.

Sundman's inequality is a special case

I now show that the Sundman inequality and my extension in Thm. 2.9 are, essentially, the inequalities obtained by dropping the $\mathbf{W}_{config}$ terms from the velocity decomposition. As such, anticipate that equality (where $\mathbf{W}_{config} = \mathbf{0}$) of these expressions is accompanied with central configurations. The proof of this theorem starts our description (Eq. 2.64) about the structure of $\mathbf{W}_{config}$.

Theorem 2.10 *The Sundman inequality, Eq. 2.60, and the Eq. 2.63 inequalities are special cases of the velocity decomposition and Eq. 2.43. If $c^2 + (I')^2 = 4IT$, then the particles are in the invariable plane (i.e., $\mathbf{Z} = \mathbf{0}$) and $\mathbf{W}_{config} = \mathbf{0}$. If $c^2 + (I')^2 \equiv 4IT$, the motion is coplanar and the homographic motion defines a central configuration.*

Proof: To prove the Sundman inequality assertion, just drop the $\mathbf{W}^2_{config}$ terms from Eq. 2.43 to get $\mathbf{W}^2_{rot} + \mathbf{W}^2_{scal} \leq \mathbf{V}^2 = 2(U + h)$ (equality occurs iff $\mathbf{W}_{config} = \mathbf{0}$), replace $\mathbf{W}^2_{rot}$ with its lower bound (Eq. 2.42) of $\frac{c^2}{R^2} \leq \frac{c^2}{R^2 - Z_2} \leq \mathbf{W}^2_{rot}$ (equality requires $\mathbf{Z} = \mathbf{0}$, i.e., all of the particles are in the invariable plane), and use $\mathbf{W}^2_{scal} = (R')^2$ to obtain

$$\frac{c^2 + (RR')^2}{2R} - hR \leq RU.$$

As $R^2 = 2I$ and $R'R = I'$, Eq. 2.60 follows. If $c^2 + (I')^2 \equiv 4IT$, then $\mathbf{W}_{config} \equiv \mathbf{0}$, so the central configuration conclusion follows from Thm. 2.7.

To show that the stronger inequalities of Eq. 2.63 are special cases of the velocity decomposition, divide $\mathbf{W}_{config}$ into

$$\mathbf{W}_{config} = \mathbf{W}_{config,rot} + \mathbf{W}_{config,scal}. \tag{2.64}$$

To introduce the terms, notice that $\mathbf{s} \times \mathbf{r}_j$ and $\frac{R'}{R}\mathbf{r}_j$ are, respectively, the jth particle's system rotational and scalar velocity components. According to Eq. 2.28, the terms $[\frac{r'_j}{r_j} - \frac{R'}{R}]\mathbf{r}_j$ and $\frac{\mathbf{r}_j \times \mathbf{v}_J}{r_j} - \mathbf{s} \times \mathbf{r}_j$ are the left-overs: they describe how the velocity of the jth particle helps to change the shape of the configuration, respectively, in the scalar and rotational directions. Definitions for the two $\mathbf{W}_{config}$ components (that obviously are orthogonal) are immediate:

$$\mathbf{W}_{config,scal} := ([\frac{r'_1}{r_1} - \frac{R'}{R}]\mathbf{r}_1, \ldots, [\frac{r'_N}{r_N} - \frac{R'}{R}]\mathbf{r}_N), \tag{2.65}$$

is the system's *scalar configurational velocity*; the system's *rotational configurational velocity* is

$$\mathbf{W}_{config,rot} := (\frac{\mathbf{r}_1 \times \mathbf{v}_1}{r_1} - \mathbf{s} \times \mathbf{r}_1, \ldots, \frac{\mathbf{r}_N \times \mathbf{v}_N}{r_N} - \mathbf{s} \times \mathbf{r}_N). \tag{2.66}$$

Because $\mathbf{W}_{scal} + \mathbf{W}_{config,scal} = (r'_1\frac{\mathbf{r}_1}{r_1}, \ldots, r'_N\frac{\mathbf{r}_N}{r_N})$ and $\mathbf{W}_{scal}, \mathbf{W}_{config,scal}$ are obviously orthogonal,[15] it follows that

$$\mathbf{W}^2_{scal} + \mathbf{W}^2_{config,scal} = 2Q. \tag{2.67}$$

The first inequality of Eq. 2.63 follows by dropping $\mathbf{W}^2_{config,scal}$ (so equality holds iff this term is zero) and using $\mathbf{W}^2_{scal} = (R')^2$ and $R^2 = 2I$. The second inequality follows in the same way by using the equality

$$\mathbf{W}_{rot} + \mathbf{W}_{config.rot} = (\frac{\mathbf{r}_1 \times \mathbf{v}_1}{r_1}, \ldots, \frac{\mathbf{r}_N \times \mathbf{v}_N}{r_N}).$$

Equality requires that $\mathbf{W}_{config,rot} = \mathbf{0}$ and coplanar motion. $\square$

---

[15]This is because $\mathbf{W}_{scal}$ is the projection of $\mathbf{V}$ in the $\mathbf{R}/R$ direction.

2.4 More conjectures

A theme of this chapter is to find natural settings where the configuration must be a central configuration, or tend to a central configuration. This suggests a natural research program that I have not systematically explored: find all situations that require $\mathbf{W}_{config} \equiv \mathbf{0}$ or $\mathbf{W}_{config}, \mathbf{W}'_{config} \to \mathbf{0}$ and for central configurations to arise. Examples are shown above,; e.g., if the motion stays in the same Euler equivalence class, or if the Sundman inequality always is an equality, the associated configuration must be a central configuration.

As a suggestion of what might be investigated for the three body problem, notice that the three particles always define a plane called the *osculating plane*. If the inclination of this plane is a constant different from zero, must the configuration be a central configuration? Cabral [8] showed this is not the case for an inclination of $\pi/2$: the motion is a rotating isosceles triangle where one particle slides up and down the z-axis. While this choice of a relationship does not work, are there others that will?

2.4.1 Another conjecture

One choice for a conjecture, which is a natural extension of motion staying in an Euler similarity class, is to examine the motion that preserves a cruder measure of the configurations. This leads to the following conjecture:

> **Conjecture:** If the configurational measure IU^2 (or RU) is a constant, then $\mathbf{W}_{config} \equiv \mathbf{0}$ and the configuration formed by the particles is a central configuration.

This conjecture trivially includes the one on page 47: if I and U are equal to constants, then so is IU^2. However, IU^2 can be a constant with varying I and U values. Indeed, according to Thm. 2.6, examples can be created by selecting a central configuration and setting $\mathbf{V} = \mathbf{0}$: the configuration is preserved while all particles head for a complete collapse of the system. To find other coplanar examples, select initial conditions $\mathbf{W}_{config} = \mathbf{0}$ for a central configuration.

To illustrate how this conjecture fits into our understanding of the dynamics of the N-body problem, recall that the "central configuration" theme of this chapter was introduced by searching for N-body solutions and settings that resemble the central force law expression $\mathbf{r}'' = \lambda \mathbf{r}$ where the force and acceleration vectors are on the same line. This discussion was captured by Eq. 2.7, which can be expressed as $\mathbf{R}'' = \lambda \mathbf{R}$ or $\lambda \mathbf{R} = \nabla_s U$. Rather than the vector equation, we could examine when the N-body dynamics of R resembles that of r. To be more specific, as

$$r'' = -\frac{A}{r^2} + \frac{c^2}{r^3},$$

the goal is to find the N-body motions where a solution satisfies

$$R'' = -\frac{A}{R^2} + \frac{B}{R^3} \qquad (2.68)$$

for some positive constants A and B. According to the proof of Thm. 2.6, this behavior occurs whenever the motion preserves a central configuration. The next result states that Eq. 2.68 holds iff RU is a constant.

Theorem 2.11 *The equations of motion for R in the coplanar N-body problem can be expressed in the Eq. 2.68 form iff RU is a constant.*

A solution for the N-body problem satisfies $R'' = -\frac{A}{R^2}$ for some positive constant A iff $c = 0$ with homothetic motion that preserves a central configuration.

Proof: As $\mathbf{R}'' = -\frac{U}{R^2}\mathbf{R} + \sum \alpha_j \mathbf{U}_j$ and $R'' = \frac{\mathbf{R}}{R} \cdot \mathbf{R}'' + \frac{\mathbf{W}_{rot}^2 + \mathbf{W}_{config}^2}{R}$, because $\mathbf{R}$ is orthogonal to the $\mathbf{U}_j$ terms, we have that

$$R'' = -\frac{(RU)}{R^2} + \frac{c^2}{R^3} + \frac{(R\mathbf{W}_{config})^2}{R^3}. \qquad (2.69)$$

If RU is a constant, then the first term on the right-hand side of Eq. 2.69 is $\frac{A}{R^2}$. Moreover, according to Thm. 2.4 , if RU is a constant, then $(R\mathbf{W}_{config})^2$ also is a constant, so Eq. 2.68 follows.

To prove the converse, if there is a solution where Eq. 2.68 holds, then by multiplying Eqs. 2.68, 2.69 by R^3, we obtain

$$-R(RU) + c^2 + (R\mathbf{W}_{config})^2 = -AR + B.$$

Differentiating this expression leads to

$$-R'(RU) + R(RU)' - [2R(RU)' - ((R\mathbf{W}_{config})^2)'] = -AR'$$

According to Thm. 2.4, the term in the bracket on the left-hand side of this expression is zero, so, by carrying out the full differentiation, we have that $U' = -\frac{A}{R^2}R'$. Thus, for this solution, $U = \frac{A}{R} + D$. If $D = 0$, ,the configurational measure RU is a constant. If $D \neq 0$, then Eq. 2.69 has a $\frac{D}{R}$

term, which contradicts the assumption that the solution has an Eq. 2.68 format. As $D = 0$, RU is a constant: this completes the first part of the proof.

The proof for $R'' = -\frac{A}{R^2}$ is done in the same way up to where $(R\mathbf{W}_{config})^2$ must be a constant. Thus $R'' = -\frac{A}{R^2} + \frac{c^2 + (R\mathbf{W}_{config})^2}{R^3}$, so the assumed form for R'' forces $c^2 + (R\mathbf{W}_{config})^2 = 0$, or $c = W_{config} = 0$. The solution now follows from the earlier results. For the converse, it is easy to show that a homothetic solution preserving a central configuration satisfies $R'' = -\frac{A}{R^2}$. $\square$

Theorem 2.11 proves the first part of a related conjecture that

> **Conjecture:** Eq. 2.68 holds iff RU is a constant iff the configuration defined by the particles always is a central configuration.

Analyzing these conjectures probably involves Eq. 2.43. After all, the goal is to prove that $\mathbf{W}_{config} = \mathbf{0}$, and $\mathbf{W}_{config}$ is in the Eq. 2.43 equality. To illustrate, I use Eq. 2.43 to prove Thm. 2.2.

Proof of Thm. 2.2: If $I = \frac{1}{2}R^2$ is a constant, then $U \equiv -2h$ so $h < 0$. According to Eq. 2.43 and the $\mathbf{W}_{rot}^2 \geq \frac{c^2}{R^2}$ lower bound, we have that

$$\frac{c^2}{R^2} < \mathbf{W}_{rot}^2 < 2(U + h) = 2|h|.$$

The first conclusion follows from this inequality as

$$\frac{c^2}{2|h|} < R^2 = 2I.$$

If the conjecture is true, then $\mathbf{W}_{scal} \equiv \mathbf{0}$ (as I is a constant) and $\mathbf{W}_{config} \equiv \mathbf{0}$ (as the configuration does not change). According to Eq. 2.43 and Thms. 2.5, 2.7, this is true iff the motion is coplanar and

$$\frac{c^2}{R^2} \equiv \mathbf{W}_{rot}^2 \equiv 2(U + h) \equiv 2|h|$$

from which the conclusion follows. $\square$

2.4.2 Special cases

Why should we believe this "a constant RU requires homographic motion" conjecture? Partial support comes from Thm. 2.4, which was central for the proof of Thm. 2.11 asserting that if RU is a constant, then so is $(R\mathbf{W}_{config})^2$:

the goal is to show that this $(R\mathbf{W}_{config})^2$ constant is zero. For more structure, as IU^2 is homogeneous of degree zero and invariant with respect to
rotations, it follows that

$$< \nabla_s IU^2, \mathbf{V} >=< \nabla_s IU^2, \mathbf{W}_{config} > \cdot$$

But $< \nabla_s IU^2, \mathbf{W}_{config} > =< U\mathbf{R} + 2IU\nabla_s U, \mathbf{W}_{config} > .$ As $\mathbf{R}$ is orthogonal to $\mathbf{W}_{config}$, these conjectures would be true if

$$< \nabla_s U, \mathbf{W}_{config} >\equiv 0 \qquad\qquad (2.70)$$

requires the motion to define a central configuration. This approach captures
the sense of the quote on page 60.

Indeed, the Eq. 2.70 constraint can be used to show that if RU is a
constant in the three-body problem when the particles are constrained to a
straight line, then the conjectures are true.

Theorem 2.12 *For the three-body problem, assume that at each instant of
time t all particles are on a line. If RU is a constant, then the particles
always define a central configuration.*

Proof: For the three body problem, a basis for the configurational velocity,
$\{\mathbf{U}_{scal}, \mathbf{U}_{rot}\}$, can be found where the components of $\mathbf{U}_{scal}$ always are along
the line defined by the particles and each component of $\mathbf{U}_{rot}$ is orthogonal
to this line; i.e., $\mathbf{W}_{config} = a\mathbf{U}_{scal} + b\mathbf{U}_{rot}$. If $b \neq 0$, then the configuration
cannot always stay on a straight line, so $\mathbf{W}_{config} = a\mathbf{U}_{scal}$. Because the
particles are on a line, it must be that $\nabla_s U = -\frac{U}{R^2}\mathbf{R} + \alpha\mathbf{U}_{scal}$. According
to Eq. 2.70, it must be that $a\alpha = 0$. If $\alpha = 0$, then the configuration is
a central configuration. If $a = 0$, then the configuration is preserved and,
according to Thm. 2.7, the configuration is a central configuration. $\square$

In fact, if $\mathbf{c} \neq \mathbf{0}$, the central configuration assertion of Thm. 2.12 holds for
any N *without* assuming that RU is a constant. The proof of this assertion
uses properties of the coplanar problem, but it is easy to show that if the
particles always define a line, then the motion is coplanar.

Theorem 2.13 *For the N-body problem, assume that at each instant of
time t all particles are on a line. If $\mathbf{c} \neq \mathbf{0}$, then the solution is homographic
where the particles define a central configuration.*[16]

---

[16]This is the earlier mentioned theorem. As M. Santoprete, F. Diacu, and E. Perez-
Chavela use my velocity decomposition, they should have a similar result, but I expect
that their approach differs.

Proof: Let $\mathbf{u}(t)$ be a unit vector defining the line. This means that $\mathbf{r}_j(t) = \pm r_j(t)\mathbf{u}(t)$ for all j. To simplify the notation of this proof, assume that r_j has the appropriate sign. As $\mathbf{u}' = \lambda \mathbf{e}_3 \times \mathbf{u}$ for some value of λ, we have that

$$
\begin{aligned}
\mathbf{r}'_j &= r'_j\mathbf{u} + \lambda r_j\mathbf{e}_3 \times \mathbf{u} = r'_j\mathbf{u} + \lambda\mathbf{e}_3 \times \mathbf{r}_j \\
\mathbf{r}''_j &= (r''_j - \lambda^2 r_j)\mathbf{u} + [(\lambda r_j)' + \lambda r'_j]\mathbf{e}_3 \times \mathbf{u}
\end{aligned}
\tag{2.71}
$$

Using the $\mathbf{e}_3 \times \mathbf{r}_j$ term from $\mathbf{r}'_j$, it follows from $\mathbf{W}_{rot} = \frac{c}{R^2}\mathbf{E}_3 \times \mathbf{R}$ that $\lambda = \frac{c}{R^2}$. As $\mathbf{r}''_j = \frac{1}{m_j}\frac{\partial U}{\partial \mathbf{r}_j}$ where the particles are on the line defined by $\mathbf{u}$, it must be that $\frac{\partial U}{\partial \mathbf{r}_j}$ is a multiple of $\mathbf{u}$. Thus, for each j, it must be that $[(\lambda r_j)' + \lambda r'_j] - 2c[r'_j\frac{1}{R^2} - r_j\frac{R'}{R^3}] = 0$. So for each j we have that $r'_j = \frac{R'}{R}r_j$, or that $\mathbf{V} = \mathbf{W}_{rot} + \mathbf{W}_{scal}$. As this representation of $\mathbf{V}$ requires $\mathbf{W}_{config} = \mathbf{0}$, the conclusion follows from Thm. 2.5. □

For a different explanation of this result, that involves the dynamics, see the discussion on page 79.

2.5   Jacobi coordinates help "see" the dynamics

Much of our discussion has involved that shadowy, mysterious $\mathbf{W}_{config}$. A minimal way to shed light on this term is to describe a basis for the spaces of configurational velocity and $\nabla_s U$. The value of doing so is obvious: the extra information will lead to new results. It does, but I only indicate how to find some new conclusions to avoid diverting attention from the goals of this book.[17] While a detailed discussion is left for elsewhere (Saari [90]), treat this basis for the coplanar N-body problem as providing closure for this velocity decomposition discussion.

To partially motivate what follows, wouldn't it be nice if we had at least a rough indication of what will happen with the dynamics of the system by just examining the shape of the current configuration? As a challenge, with the positioning of the three particles in Fig. 2.3a: what do you think will happen next? The answer clearly depends upon the current value of $\mathbf{V}$, so what do you think will happen in special cases such as where the masses are all equal and $\mathbf{W}_{config} = \mathbf{0}$? As I indicate next, we can obtain insight into such problems by using a basis for the configurational velocity and $\nabla_s U$. The approach is to express both $\nabla_s U$ and $\mathbf{R}''$ in terms of the basis: this establishes a natural connection—a bridge—between the present configuration and how the velocity—and the configuration—must change.

---

[17]The purpose of this book is to report on the CMBS lectures.

A convenient way to introduce the basis $\{\mathbf{U}_j\}$ for the configurational velocity and for $\nabla_s U$ is to first discuss the three-body problem where the space has dimension two. Our introduction of $\mathbf{W}_{config,scal}$ and $\mathbf{W}_{config,rot}$ (page 64) suggests using the vectors $\mathbf{U}_{scal}$ and $\mathbf{U}_{rot}$: $\mathbf{U}_{scal}$ captures $\mathbf{V}$ terms in the $\frac{\mathbf{r}_j}{r_j}$ directions that are *not* extracted by $\mathbf{W}_{scal}$, and $\mathbf{U}_{rot}$ does the same for the $\mathbf{V}$ rotational terms that are *not* extracted by $\mathbf{W}_{rot}$. A difficulty in finding an explicit form for these vectors comes from the need to satisfy the center of mass conditions. To avoid these complications, it is convenient to describe $\mathbf{U}_{rot}$ and $\mathbf{U}_{scal}$ with Jacobi coordinates.

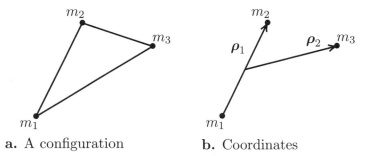

a. A configuration   **b.** Coordinates

Fig. 2.3. Jacobi coordinates

An advantage of using Jacobi coordinates is that they reduce the system by incorporating the center of mass integrals (Eqs. 2.2, 2.36) into the equations of motion. Roughly speaking, think of the approach as describing the three-body problem as two two-body problems. The first "two-body problem" is the position of m_2 relative to m_1. The other "two-body problem" expresses the location of the third particle relative to the center of mass, $\frac{m_1\mathbf{r}_1+m_2\mathbf{r}_2}{m_1+m_2}$, of the first two particles. The new coordinates, as depicted in Fig. 2.3b, are

$$\boldsymbol{\rho}_1 = \mathbf{r}_2 - \mathbf{r}_1, \quad \boldsymbol{\rho}_2 = \mathbf{r}_3 - \frac{m_1\mathbf{r}_1 + m_2\mathbf{r}_2}{m_1 + m_2}. \tag{2.72}$$

To find the equations of motion, we need the representations for the different $\mathbf{r}_j - \mathbf{r}_k$ terms. By use of elementary vector analysis, they are

$$\mathbf{r}_2 - \mathbf{r}_1 = \boldsymbol{\rho}_1, \quad \mathbf{r}_3 - \mathbf{r}_1 = \boldsymbol{\rho}_2 + \frac{m_2}{m_1 + m_2}\boldsymbol{\rho}_1, \quad \mathbf{r}_3 - \mathbf{r}_2 = \boldsymbol{\rho}_2 - \frac{m_1}{m_1 + m_2}\boldsymbol{\rho}_1,$$

along with $\boldsymbol{\rho}_2 = \frac{m_1+m_2+m_3}{m_1+m_2}\mathbf{r}_3$.

Substituting these values into $\boldsymbol{\rho}_1'' = \mathbf{r}_2'' - \mathbf{r}_1''$ and $\boldsymbol{\rho}_2'' = \frac{m_1+m_2+m_3}{m_1+m_2}\mathbf{r}_3''$ yields

$$\begin{aligned}
\boldsymbol{\rho}_1'' &= -\left(\frac{m_1+m_2}{\rho_1^3} + \frac{m_3}{m_1+m_2}\{\frac{m_1}{r_{2,3}^3} + \frac{m_2}{r_{1.3}^3}\}\right)\boldsymbol{\rho}_1 - m_3\left(\frac{1}{r_{1,3}^3} - \frac{1}{r_{2,3}^3}\right)\boldsymbol{\rho}_2 \\
\boldsymbol{\rho}_2'' &= -\frac{m_3\mu_1}{\mu_2}\left(\frac{1}{r_{13}^3} - \frac{1}{r_{2,3}^3}\right)\boldsymbol{\rho}_1 - \frac{m_3}{\mu_2}\left(\frac{m_1}{r_{1,3}^3} + \frac{m_2}{r_{2,3}^3}\right)\boldsymbol{\rho}_2
\end{aligned} \tag{2.73}$$

where the generalized masses are

$$\mu_1 = \frac{m_1 m_2}{m_1 + m_2}, \quad \mu_2 = \frac{m_3(m_1 + m_2)}{m_1 + m_2 + m_3}. \tag{2.74}$$

Let the new system inner product for $\mathbf{a} = (\mathbf{a}_1, \mathbf{a}_2), \mathbf{b} = (\mathbf{b}_1, \mathbf{b}_2) \in (\mathbb{R}^2)^3$ be

$$[\mathbf{a}, \mathbf{b}] = \sum_{j=1}^{2} \mu_j \mathbf{a}_j \cdot \mathbf{b}_j, \tag{2.75}$$

Eq. 2.73 becomes

$$\boldsymbol{\rho}'' = (\boldsymbol{\rho}_1'', \boldsymbol{\rho}_2'') = \nabla_J U \tag{2.76}$$

where ∇_J represents the gradient with respect to these Jacobi coordinates and the $[,]$ scalar product.

2.5.1 Velocity decomposition and a basis

For $\boldsymbol{\rho} = (\boldsymbol{\rho}_1, \boldsymbol{\rho}_2)$, the velocity decomposition is

$$\mathbf{W}_{rot} = \mathbf{S} \times \boldsymbol{\rho}, \quad \mathbf{W}_{scal} = \rho' \frac{\boldsymbol{\rho}}{\rho}, \quad \mathbf{W}_{config} = \boldsymbol{\rho}' - (\mathbf{W}_{rot} + \mathbf{W}_{scal}) \tag{2.77}$$

while

$$\begin{aligned}
\mathbf{W}_{config,rot} &= (\frac{\boldsymbol{\rho}_1 \times \boldsymbol{\rho}_1'}{\rho_1} - \mathbf{s} \times \boldsymbol{\rho}_1, \frac{\boldsymbol{\rho}_2 \times \boldsymbol{\rho}_2'}{\rho_2} - \mathbf{s} \times \boldsymbol{\rho}_2) \\
\mathbf{W}_{config,scal} &= ([\frac{\rho_1'}{\rho_1} - \frac{\rho'}{\rho}]\boldsymbol{\rho}_1, [\frac{\rho_2'}{\rho_2} - \frac{\rho'}{\rho}]\boldsymbol{\rho}_2).
\end{aligned} \tag{2.78}$$

Theorem 2.14 *For the coplanar three-body problem expressed in Jacobi coordinates, the following is a basis for the configurational velocity:*

$$\mathbf{U}_{scal} = (\frac{\rho_2}{\mu_1}(\frac{\boldsymbol{\rho}_1}{\rho_1}), -\frac{\rho_1}{\mu_2}(\frac{\boldsymbol{\rho}_2}{\rho_2})), \quad \mathbf{U}_{rot} = (\frac{\rho_2}{\mu_1}(\mathbf{e}_3 \times \frac{\boldsymbol{\rho}_1}{\rho_1}), -\frac{\rho_1}{\mu_2}(\mathbf{e}_3 \times \frac{\boldsymbol{\rho}_2}{\rho_2})). \tag{2.79}$$

Namely, at each time t there exist constants a and b so that

$$\mathbf{W}_{config,scal} = a\mathbf{U}_{scal}, \quad \mathbf{W}_{config,rot} = b\mathbf{U}_{rot} = b\mathbf{E}_3 \times \mathbf{U}_{scal}. \tag{2.80}$$

Because there are three choices of selecting which two particles will define $\boldsymbol{\rho}_1$ (actually, with direction, six choices) to create the Jacobi coordinates, the several corresponding choices for this $\{\mathbf{U}_{scal}, \mathbf{U}_{rot}\}$ basis adds flexibility in analyzing behavior. This will be briefly mentioned later.

Proof: Because $\mathbf{U}_{rot} = \mathbf{E}_3 \times \mathbf{U}_{scal}$, it is clear that $\mathbf{U}_{scal}$ and $\mathbf{U}_{rot}$ are orthogonal. To be a basis for the space of configurational velocity, $\mathbf{U}_{scal}$

and $\mathbf{U}_{rot}$ must also be orthogonal to $\boldsymbol{\rho}$ and $\mathbf{E}_3 \times \boldsymbol{\rho}$: this involves a simple computation. (These computations explain the role of the $\frac{1}{\mu_j}$ terms.) $\square$

To find a basis for the general three-body problem, replace $\mathbf{e}_3$ in Eq. 2.79 with a unit vector normal to the plane defined by the three bodies. That is, replace $\mathbf{e}_3$ with a unit vector along the line $\boldsymbol{\rho}_1 \times \boldsymbol{\rho}_2$: my preferred choice (to avoid sign problems) is the one that has a positive dot product with $\mathbf{s}$.

These $\{\mathbf{U}_j\}$ choices make sense. The radial velocity $\mathbf{W}_{scal}$, for instance, extracts as much as possible from $\boldsymbol{\rho}'$ that is in the $(\rho_1(\frac{\boldsymbol{\rho}_1}{\rho_1}), \rho_2(\frac{\boldsymbol{\rho}_2}{\rho_2}))$ direction. Any remaining terms, then, have the $(a\frac{\boldsymbol{\rho}_1}{\rho_1}, b\frac{\boldsymbol{\rho}_2}{\rho_2})$ form where a and b have different signs. The actual coefficients come from computations.

Properties of the basis

Some useful and immediate properties of this basis follow:

Theorem 2.15 *For the coplanar three-body problem,*

$$\mathbf{U}_{rot}^2 = \mathbf{U}_{scal}^2 = \frac{\rho^2}{\mu_1 \mu_2}. \tag{2.81}$$

A relationship among these two orthogonal vectors is

$$\mathbf{E}_3 \times \mathbf{U}_{rot} = -\mathbf{U}_{scal}, \quad \mathbf{E}_3 \times \mathbf{U}_{scal} = \mathbf{U}_{rot}. \tag{2.82}$$

Defining the rotational rate λ_j by $(\frac{\boldsymbol{\rho}_j}{\rho_j})' = \lambda_j \mathbf{e}_3 \times \frac{\boldsymbol{\rho}_j}{\rho_j}$, and assuming that $\mathbf{W}_{config} = \mathbf{W}_{config,scal} + \mathbf{W}_{config,rot} = a\mathbf{U}_{scal} + b\mathbf{U}_{rot}$, we have

$$a\mathbf{U}_{scal}^2 = \rho_1' \rho_2 - \rho_2' \rho_1, \quad b\mathbf{U}_{rot}^2 = (\lambda_1 - \lambda_2)\rho_1 \rho_2, \tag{2.83}$$

Thus,

$$\mathbf{W}_{config,scal}^2 = a(\frac{\rho_1'}{\rho_1} - \frac{\rho_2'}{\rho_2})\rho_1 \rho_2, \quad \mathbf{W}_{config,rot}^2 = b(\lambda_1 - \lambda_2)\rho_1 \rho_2 \tag{2.84}$$

and the a, b coefficients have the values

$$a = \frac{\mu_1 \mu_2 (\rho_1' \rho_2 - \rho_2' \rho_1)}{\rho^2}, \quad b = \frac{\mu_1 \mu_2 \rho_1 \rho_2 (\lambda_1 - \lambda_2)}{\rho^2}. \tag{2.85}$$

Expressing $\mathbf{U}'_{scal}$ and $\mathbf{U}'_{rot}$ in terms of the basis yields

$$\mathbf{U}'_{scal} = \frac{(\lambda_1 - \lambda_2)\rho_1 \rho_2}{\rho^2} \mathbf{E}_3 \times \boldsymbol{\rho} + \frac{\rho_1 \rho_2' - \rho_2 \rho_1'}{\rho^2} \boldsymbol{\rho} + \frac{(\rho^2)'}{2\rho^2} \mathbf{U}_{scal} + \frac{\rho_\lambda^2}{\rho^2} \mathbf{U}_{rot} \tag{2.86}$$

$$\mathbf{U}'_{rot} = \frac{(\lambda_2 - \lambda_1)\rho_1\rho_2}{\rho^2}\boldsymbol{\rho} + \frac{\rho_1\rho'_2 - \rho_2\rho'_1}{\rho^2}\mathbf{E}_3 \times \boldsymbol{\rho} + \frac{(\rho^2)'}{2\rho^2}\mathbf{U}_{rot} - \frac{\rho_\lambda^2}{\rho^2}\mathbf{U}_{scal} \quad (2.87)$$

where

$$\rho_\lambda^2 := \lambda_2\mu_1\rho_1^2 + \lambda_1\mu_2\rho_2^2. \quad (2.88)$$

Furthermore, $\rho_\lambda^2 = c$ and $\lambda_1 = \lambda_2 = \frac{c}{\rho^2}$, iff $\mathbf{W}_{config,rot} = \mathbf{0}$.

Observe that λ_j is related to $\boldsymbol{\rho}_j$'s angular velocity: i.e., if $\boldsymbol{\rho}'_j = \alpha_j\rho'_j\frac{\boldsymbol{\rho}_j}{\rho_j} + \beta_j\mathbf{e}_3\times\frac{\boldsymbol{\rho}_j}{\rho_j}$, then $(\frac{\boldsymbol{\rho}_j}{\rho_j})' = \frac{\beta_j}{\rho_j}\mathbf{e}_3\times\frac{\boldsymbol{\rho}_j}{\rho_j}$, or $\lambda_j\rho_j = \beta_j = \mathbf{e}_3\cdot(\frac{\boldsymbol{\rho}_j}{\rho_j}\times\boldsymbol{\rho}'_j)$. (I use λ_j rather than β_j to suggest comparisons with $\mathbf{W}_{rot}$.) Notice the symmetry in the coefficients of Eqs. 2.86, 2.87; e.g., what the $(\lambda_1 - \lambda_2)\rho_1\rho_2 = (\frac{\beta_1}{\rho_1} - \frac{\beta_2}{\rho_2})\rho_1\rho_2$ term does for differences in angular changes is what the $\rho_2\rho'_1 - \rho_1\rho'_2 = (\frac{\rho'_1}{\rho_1} - \frac{\rho'_2}{\rho_2})\rho_1\rho_2$ term does for differences in radial changes.

Proof: The proof of Eq. 2.82 is immediate. For the proof of Eq. 2.81,

$$\mathbf{U}^2_{scal} = \mu_1[\frac{\rho_2}{\mu_1}]^2 + \mu_2[\frac{\rho_1}{\mu_2}]^2 = \frac{\rho_2^2}{\mu_1} + \frac{\rho_1^2}{\mu_2} = \frac{\rho^2}{\mu_1\mu_2},$$

where a similar computation holds for $\mathbf{U}^2_{rot}$.

To prove the first part of Eq. 2.83, because $\boldsymbol{\rho}$ is orthogonal to the space containing $\mathbf{W}_{config}$, we have that $[\boldsymbol{\rho}, \mathbf{U}_{scal}] = 0$. Differentiating leads to

$$[\boldsymbol{\rho}', \mathbf{U}_{scal}] + [\boldsymbol{\rho}, \mathbf{U}'_{scal}] = 0 \quad (2.89)$$

where, according to the velocity decomposition

$$\boldsymbol{\rho}' = \mathbf{W}_{rot} + \mathbf{W}_{scal} + a\mathbf{U}_{scal} + b\mathbf{U}_{rot},$$

the first terms equals $a\mathbf{U}^2_{scal}$. To compute the second term,

$$\mathbf{U}'_{scal} = (\frac{\rho'_2}{\mu_1}(\frac{\boldsymbol{\rho}_1}{\rho_1}), -\frac{\rho'_1}{\mu_2}(\frac{\boldsymbol{\rho}_2}{\rho_2})) + (\frac{\rho_2}{\mu_1}(\frac{\boldsymbol{\rho}_1}{\rho_1})', -\frac{\rho_1}{\mu_2}(\frac{\boldsymbol{\rho}_2}{\rho_2})').$$

Because $\frac{\boldsymbol{\rho}_j}{\rho_j}$, $j = 1, 2$, is a unit vector, $(\frac{\boldsymbol{\rho}_j}{\rho_j})' = \lambda_j\mathbf{e}_3\times\frac{\boldsymbol{\rho}_j}{\rho_j}$ for an appropriate λ_j value, and $\mathbf{e}_3 \times \boldsymbol{\rho}_j$ is orthogonal to $\boldsymbol{\rho}_j$. Thus

$$[\boldsymbol{\rho}, \mathbf{U}'_{scal}] = \mu_1\frac{\rho'_2}{\mu_1}[\boldsymbol{\rho}_1 \cdot \frac{\boldsymbol{\rho}_1}{\rho_1}] - \mu_2\frac{\rho'_1}{\mu_2}[\boldsymbol{\rho}_2 \cdot \frac{\boldsymbol{\rho}_2}{\rho_2}] = \rho'_2\rho_1 - \rho'_1\rho_2.$$

The conclusion follows by substituting the computed values into Eq. 2.89.

The same argument leads to $[\boldsymbol{\rho}', \mathbf{U}_{rot}] + [\boldsymbol{\rho}, \mathbf{U}'_{rot}] = b\mathbf{U}^2_{rot} + [\boldsymbol{\rho}, \mathbf{U}'_{rot}] = 0$. But

$$\mathbf{U}'_{rot} = (\frac{\rho'_2}{\mu_1}(\mathbf{e}_3 \times \frac{\boldsymbol{\rho}_1}{\rho_1}), -\frac{\rho'_1}{\mu_2}(\mathbf{e}_3 \times \frac{\boldsymbol{\rho}_2}{\rho_2})) + (\frac{\rho_2}{\mu_1}(-\lambda_1\frac{\boldsymbol{\rho}_1}{\rho_1}), -\frac{\rho_1}{\mu_2}(-\lambda_2\frac{\boldsymbol{\rho}_2}{\rho_2})).$$

Equations 2.84 and 2.85 follow from $\mathbf{W}^2_{config,scal} = a^2\mathbf{U}^2_{scal}$ and $\mathbf{W}^2_{config,rot} = b^2\mathbf{U}^2_{rot}$.

To derive Eq. 2.86, find the values $[\mathbf{U}'_{scal}, \frac{\boldsymbol{\rho}}{\rho}]$, $[\mathbf{U}'_{scal}, \frac{\mathbf{E}_3 \times \boldsymbol{\rho}}{\rho}]$, $[\mathbf{U}'_{scal}, \frac{\mathbf{U}_{scal}}{U_{scal}}]$, and $[\mathbf{U}'_{scal}, \frac{\mathbf{U}_{rot}}{U_{rot}}]$. These straightforward computations use earlier values computed in this proof. The verification of Eq. 2.87 is done is a similar manner. The similarity in the two equations is to be expected from the $\mathbf{U}_{rot} = \mathbf{E}_3 \times \mathbf{U}_{scal}$ expression.

To prove the last sentence of the theorem, obviously, $\rho_\lambda^2 = \lambda\rho^2$ for some λ iff $\lambda_1 = \lambda_2$, and this common value is λ. According to Eq. 2.83, $\mathbf{W}^2_{config,rot} = b^2\mathbf{U}^2_{rot} = b\rho_1\rho_2(\lambda_1 - \lambda_2)$, so if $\lambda_1 = \lambda_2$ then $\mathbf{W}_{config,rot} = \mathbf{0}$. To proved the converse and find λ, according to Eq. 2.78,

$$\mathbf{W}^2_{config,rot} = \sum_{j=1}^2 \mu_j[\frac{\boldsymbol{\rho}_j \times \boldsymbol{\rho}'_j}{\rho_j} - \mathbf{s} \times \boldsymbol{\rho}_j]^2 = \sum_{j=1}^2 \mu_j[(\lambda_j\rho_j - \frac{c\rho_j}{\rho^2})\mathbf{e}_3 \times \frac{\boldsymbol{\rho}_j}{\rho_j}]^2$$

Thus $\mathbf{W}^2_{config,rot} = 0$ iff $\lambda_1 = \lambda_2 = \frac{c}{\rho^2}$. The conclusion now follows. □

2.5.2   Describing ρ'' with the basis

The value of Thm. 2.15 and Eqs. 2.87, 2.86 is that they provide the needed structure to express $\boldsymbol{\rho}''$ in terms of the basis. These results allow us to take a big step toward understanding how the shape of the configuration affects the dynamics.

Theorem 2.16 *Expressing* $\mathbf{W}_{config} = a\mathbf{U}_{scal} + b\mathbf{U}_{rot}$, *the* $\boldsymbol{\rho}''$ *term for the coplanar three-body problem is*

$$\boldsymbol{\rho}'' = [\frac{\rho''}{\rho} - \frac{c^2 + (\rho\mathbf{W}_{config})^2}{\rho^4}]\boldsymbol{\rho} \quad +(\frac{(a\rho^2)' - b(c + \rho_\lambda^2)}{\rho^2})\mathbf{U}_{scal} \qquad (2.90)$$
$$+(\frac{(b\rho^2)' + a(c + \rho_\lambda^2)}{\rho^2})\mathbf{U}_{rot}$$

Proof: Using the computations from Thm. 2.15 with $\boldsymbol{\rho}'' = \mathbf{W}'_{rot} + \mathbf{W}'_{scal} + (a\mathbf{U}_{scal} + b\mathbf{U}_{rot})'$, we have

$$\boldsymbol{\rho}'' = [\frac{\rho''}{\rho} - \frac{c^2}{\rho^4}]\boldsymbol{\rho} + \frac{c}{\rho^2}\mathbf{E}_3 \times \mathbf{W}_{config} + \frac{\rho'}{\rho}\mathbf{W}_{config} + (a\mathbf{U}_{scal} + b\mathbf{U}_{rot})'$$

or

$$\begin{aligned}
\boldsymbol{\rho}'' &= [\tfrac{\rho''}{\rho} - \tfrac{c^2}{\rho^4}]\boldsymbol{\rho} + \tfrac{a(\rho_1\rho_2' - \rho_2\rho_1') + b(\lambda_2 - \lambda_1)\rho_1\rho_2}{\rho^2}\boldsymbol{\rho} + \tfrac{b(\rho_1\rho_2' - \rho_2\rho_1') + a(\lambda_1 - \lambda_2)\rho_1\rho_2}{\rho^2}\mathbf{E}_3 \times \boldsymbol{\rho} \\
&\quad + (-\tfrac{cb}{\rho^2} + \tfrac{a\rho'}{\rho} + a' + \tfrac{a(\rho^2)' - 2b(\rho_\lambda^2)}{2\rho^2})\mathbf{U}_{scal} + (\tfrac{ca}{\rho^2} + \tfrac{b\rho'}{\rho} + b' + \tfrac{b(\rho^2)' + 2a\rho_\lambda^2}{2\rho^2})\mathbf{U}_{rot} \\
&= [\tfrac{\rho''}{\rho} - \tfrac{c^2}{\rho^4}]\boldsymbol{\rho} - \tfrac{\mathbf{W}_{config}^2}{\rho^2}\boldsymbol{\rho} + (-\tfrac{cb}{\rho^2} + \tfrac{a\rho'}{\rho} + a' + \tfrac{a(\rho^2)' - 2b\rho_\lambda^2}{2\rho^2})\mathbf{U}_{scal} \\
&\quad + (\tfrac{ca}{\rho^2} + \tfrac{b\rho'}{\rho} + b' + \tfrac{b(\rho^2)' + 2a\rho_\lambda^2}{2\rho^2})\mathbf{U}_{rot} \\
&= [\tfrac{\rho''}{\rho} - \tfrac{c^2 + (\rho\mathbf{W}_{config})^2}{\rho^4}]\boldsymbol{\rho} + (-\tfrac{cb}{\rho^2} + 2\tfrac{a\rho'}{\rho} + a' - \tfrac{b\rho_\lambda^2}{\rho^2})\mathbf{U}_{scal} \\
&\quad + (\tfrac{ca}{\rho^2} + 2\tfrac{b\rho'}{\rho} + b' + \tfrac{a\rho_\lambda^2}{\rho^2})\mathbf{U}_{rot} \\
&= [\tfrac{\rho''}{\rho} - \tfrac{c^2 + (\rho\mathbf{W}_{config})^2}{\rho^4}]\boldsymbol{\rho} + (\tfrac{(a\rho^2)' - b(c + \rho_\lambda^2)}{\rho^2})\mathbf{U}_{scal} + (\tfrac{(b\rho^2)' + a(c + \rho_\lambda^2)}{\rho^2})\mathbf{U}_{rot},
\end{aligned}$$

which is Eq. 2.90. In these computations, several of the equalities derived in Thm. 2.15 are used. This includes the values for $\mathbf{U}_{scal}^2 = \mathbf{U}_{rot}^2$ and the alternative representations in terms of ρ_j' and λ_j. $\square$

An immediate consequence of Eq. 2.90 is that if at some instant of time $\mathbf{W}_{config}' = \mathbf{0}$ and $\mathbf{W}_{config}' = \mathbf{0}$, then the solution defines a central configuration. This is because all of the a, a', b, b' terms equal zero. By using the basis for the general N-body problem, it turns out that the same conclusion holds. In other words, the earlier results using the $\mathbf{W}_{config} \equiv \mathbf{0}$ assumption become easier to prove.

Theorem 2.17 *For the coplanar three-body problem, if there is an instant of time where $\mathbf{W}_{config} = \mathbf{W}_{config}' = \mathbf{0}$, then $\mathbf{W}_{config} \equiv \mathbf{0}$ and the homographic solutions always preserves a central configuration.*

Proof: If $\mathbf{W}_{config} = \mathbf{W}_{config}' = \mathbf{0}$, then $a = a' = b = b' = 0$, so Eq. 2.90 becomes

$$\boldsymbol{\rho}'' = [\tfrac{\rho''}{\rho} - \tfrac{c^2}{\rho^4}]\boldsymbol{\rho} = \nabla_J U,$$

which is the equation of a central configuration.

For the coplanar problem, $\mathbf{W}_{config} = \mathbf{W}_{config}' = \mathbf{0}$ characterize the initial conditions applied to a central configuration that lead to Thm. 2.6, so, according to uniqueness results from differential equations, the conclusion follows. To explain how the result differs from the general three-body problem, if the configuration is not in the invariable plane, then $\mathbf{U}_{rot}$ is defined by a unit normal vector to the osculating plane rather than $\mathbf{e}_3$. The derivative of this vector is not zero, so while $\mathbf{W}_{config}$ may be zero, $\mathbf{W}_{config}'$ is not: the equations of motion differ. $\square$

The remaining step in the program of this section is to express $\nabla_J U$ in terms of the basis: this is done next.

2.5.3　"Seeing" the gradient of U

What makes these $\mathbf{U}_j$ terms useful is that

$$\nabla_J U = -\frac{U}{\rho^2}\rho + \alpha\mathbf{U}_{scal} + \beta\mathbf{U}_{rot}. \tag{2.91}$$

An interesting feature about Eq. 2.91 is that, by expressing $\frac{U}{\rho^2}$ in the $\frac{\rho U}{\rho^3}$ form to emphasize that this term is homogeneous of degree -3, the new expression has the configurational measure as a coefficient. This expression, then, suggests how the configurational measure affects the dynamics. (Recall from page 42 that RU has its minimum value at a regular configuration, such as an equilateral triangle, but it can have an arbitrarily large value should some particles approach each other.) Another feature of this equation is that it provides information about the configuration. For instance, if $\mathbf{W}_{config} = a\mathbf{U}_{scal} + b\mathbf{U}_{rot}$, then the earlier $[\nabla_J U, \mathbf{W}_{config}] = 0$ condition from Eq. 2.70 requires $a\alpha + b\beta = 0$.

Of particular interest, the following result, which helps us "see" when these $\mathbf{U}_j$ terms occur, simplifies the analysis.

Theorem 2.18 *For the coplanar three-body problem, $\nabla_J U$ does not have a $\mathbf{U}_{rot}$ component iff the particles form a straight line, or, for the non-collinear problem, the configuration is that of an isosceles triangle where $r_{2,3} = r_{1,3}$. If $\nabla_J U$ has a $\mathbf{U}_{rot}$ term, then the sign of β in Eq. 2.91 is the sign of $((\mathbf{e}_3 \times \rho_1) \cdot \rho_2)(r_{1,3} - r_{2,3})$. Indeed,*

$$\frac{\rho_1\rho_2}{\mu_1}\beta = -m_3\mu_1\left(\frac{1}{r_{1,3}^3} - \frac{1}{r_{2,3}^3}\right)[(\mathbf{e}_3 \times \rho_1) \cdot \rho_2]. \tag{2.92}$$

Similarly, if $\nabla_J U$ has a $\mathbf{U}_{scal}$ term, then

$$\frac{\rho^2}{\mu_1\mu_2}\alpha = \rho_1\rho_2\left(\frac{m_1}{r_{1,3}^3} + \frac{m_2}{r_{2,3}^3} - \frac{m_1+m_2}{\rho_1^3}\right) + m_3\cos(\theta)\left(\frac{1}{r_{1,3}^3} - \frac{1}{r_{2,3}^3}\right)\left(\frac{\mu_1}{\mu_2}\rho_1^2 - \rho_2^2\right) \tag{2.93}$$

where θ is the angle between ρ_1 and ρ_2.

Proof: The right-hand side of Eq. 2.73 is $\nabla_J U$, so the results follow by computing $[\nabla_J U, \mathbf{U}_{rot}]$ and $[\nabla_J U, \mathbf{U}_{scal}]$ with Eqs. 2.73 and 2.91. $\square$

Because the basis $\{\mathbf{U}_{rot}, \mathbf{U}_{scal}\}$ for the general three-body problem closely resembles that for the coplanar problem, and because this result depends only on scalar products, Thm. 2.18 holds for the general three-body problem.

The value added by Thms. 2.90, and 2.18 is that they tie together properties of the configuration and the dynamics. This is because, by equating terms on both sides of $\rho'' = \nabla_J U$, we have that

$$\alpha = \frac{(a\rho^2)' - b(c + \rho_\lambda^2)}{\rho^2}, \quad \beta = \frac{(b\rho^2)' + a(c + \rho_\lambda^2)}{\rho^2}. \tag{2.94}$$

For instance, if the current configuration is not collinear and not an isosceles triangle with $r_{1,3} = r_{2,3}$, then Eq. 2.92 ensures that $\beta \neq 0$. As a non-zero β ensures that the $\mathbf{U}_{rot}$ coefficient in ρ'' is not zero, we now know that there will be a rotational effect to change the shape of the configuration.

Recall the earlier comment made when the basis $\{\mathbf{U}_{scal}, \mathbf{U}_{rot}\}$ was introduced that each of the six choices for a Jacobi coordinate system has a de facto different basis for the configurational velocity. Consequently, we have the following result.

Corollary 2.2 *If a three-body configuration is not an equilateral triangle, then the subsequent motion will change its shape.*

Proof: If the triangle formed by the configuration is not an equilateral triangle, then there is a Jacobi coordinate system where the $\mathbf{U}_{rot}$ term is non-zero in $\nabla_J U$. The conclusion follows from the above discussion.□

2.5.4   An illustrating example

A way to illustrate Thm. 2.18 is to describe what will happen with the Fig. 2.3 configuration. First notice that $\nabla_J U$ has a $\mathbf{U}_{rot}$ term with a negative coefficient. That $\nabla_J U$ has a $\mathbf{U}_{rot}$ term follows from the fact the configuration is not an isosceles triangle. That this $\mathbf{U}_{rot}$ coefficient must be negative follows from Thm. 2.18 by observing that $r_{1,3} > r_{2,3}$ and that $(\mathbf{e}_3 \times \boldsymbol{\rho}_1) \cdot \boldsymbol{\rho}_2 < 0$; e.g., the coefficient for $\boldsymbol{\rho}_2$ in $\boldsymbol{\rho}_1''$ must be negative.

Whether $\nabla_J U$ has a $\mathbf{U}_{scal}$ component depends on the values of the masses. According to the drawing, $\rho_1 \geq r_{1,2}, r_{2,3}$, so the Eq. 2.93 term in the first parenthesis is positive. As for the second term, $\cos(\theta)$ is a small positive value and $\frac{1}{r_{1,3}^3} - \frac{1}{r_{2,3}^3} < 0$. The conclusion requires knowing the mass values: according to the figure (where the center of mass for the first two particles is at the center) $m_1 = m_2$. So, should the masses be such that $\frac{\mu_1}{\mu_2}\rho_1^2 - \rho_2^2 = \frac{m_3 + 2m_1}{4m_3}\rho_1^2 - \rho_2^2 \leq 0$ then $\nabla_J U$ has a $\mathbf{U}_{scal}$ term with a positive coefficient. As ρ_1 and ρ_2 have nearly the same length, this positive coefficient assertion would hold for $m_1 = m_2 = m_3$. More generally, a positive $\mathbf{U}_{scal}$ coefficient arises with a sufficiently large, or a sufficiently small, value of m_3. (A sufficiently small m_3 value would negate the impact of the second term in Eq. 2.93.)

Intuitive arguments and the earlier challenge

Going beyond the technicalities, often it is not overly difficult to detect what should be the signs of the two $\mathbf{U}_j$ coefficients. In Fig. 2.3, for instance, the forces acting on ρ_1 in $\nabla_J U$ clearly are pulling the head of the ρ_1 vector to the right. Thus, $\nabla_J U$ must have a negative coefficient for $\mathbf{U}_{rot}$. Detecting the sign of the coefficient for $\mathbf{U}_{scal}$ is more difficult. The idea is that the projection of $\nabla_J U$ in the $\frac{\rho}{\rho}$ direction, $-\frac{U}{\rho^2}\rho$, will be too large in one ρ_j direction and too small in the other: these differences are corrected by the $\mathbf{U}_{scal}$ term. So, the direction where the force is the greatest—for instance, where the particles are the closer to one another—is shortchanged in $\nabla_J U$: the $\alpha \mathbf{U}_{scal}$ term compensates for these differences. This is what happens with the Fig. 2.3 configuration where the force on m_3, which has components in the negative ρ_2 direction, is greater than that for ρ_1. This difference is manifested by the positive α coefficient on $\mathbf{U}_{scal}$ needed to add a negative $\frac{\rho_2}{\rho_2}$ term back to the force vector for ρ_2''.

Now consider the challenge to identify what happens to the dynamics of the Fig. 2.3 configuration with initial conditions $\mathbf{W}_{config} = \mathbf{0}$. This assumption requires $a = b = 0$ at the initial time, so initially the equations of motion are

$$
\begin{aligned}
\rho'' &= [\frac{\rho''}{\rho} - \frac{c^2}{\rho^4}]\rho + \frac{(a\rho^2)'}{\rho^2}\mathbf{U}_{scal} + \frac{(b\rho^2)'}{\rho^2}\mathbf{U}_{rot} \\
&= [\frac{\rho''}{\rho} - \frac{c^2}{\rho^4}]\rho + a'\mathbf{U}_{scal} + b'\mathbf{U}_{rot}.
\end{aligned}
$$

But the $\mathbf{U}_{scal}$ and $\mathbf{U}_{rot}$ coefficients in $\nabla_J U$ for Fig. 2.3 are, respectively, positive and negative, so initially $a' > 0$, $b' < 0$. Thus, at least in a subsequent small interval of time, $a > 0$. But as the sign of a agrees with the sign of $\frac{\rho_1'}{\rho_1} - \frac{\rho_2'}{\rho_2}$ (Eq. 2.84), this means that changes in the configurational shape will have ρ_1 larger relative to ρ_2. Similarly, b will become negative, which forces the angle between ρ_1 and ρ_2 to decrease.

It remains to compare the ρ terms from Eq. 2.90 and the $\nabla_J U$ representation for this Fig. 2.3 configuration and the assumed $\mathbf{W}_{config} = \mathbf{0}$. With these initial conditions, we have that

$$
\rho'' = -\frac{(\rho U)}{\rho^2} + \frac{c^2}{\rho^3},
$$

so ρ' is at a critical point iff the configurational measure ρU equals $\frac{c^2}{\rho}$, and ρ' increases iff the configurational measure ρU is larger than $\frac{c^2}{\rho}$.

Particles on a line

Equation 2.90 illuminates a difference between motion where the particles always define a straight line, and collinear motion where the line is fixed. The difference is that $c \neq 0$ in the first case, and $c = 0$ in the second.

In the first case, to keep the particles on a straight line, $\mathbf{W}_{config} = a\mathbf{U}_{scal}$; i.e., $b = 0$ in the $\mathbf{W}_{config}$ expression. But even should $b = 0$ in the Eq. 2.90, $\mathbf{U}_{rot}$ has the $\frac{2ac}{\rho^2}$ coefficient.[18] If this term is non-zero, then $\nabla_J U$ has a non-zero coefficient for $\mathbf{U}_{rot}$: this means that the configuration is not along a line. Thus if $c \neq 0$, it must be that $a \equiv 0$, which requires $\mathbf{W}_{config} \equiv \mathbf{0}$ and the configuration must always be a central configuration.

In contrast, the fixed position of the line for the collinear case forces $c = \lambda = 0$: these zero values make the $\mathbf{U}_{rot}$ coefficient equal to zero even if $a \neq 0$. Indeed, the equations of motion become

$$\rho'' = [\frac{\rho''}{\rho} - \frac{a^2}{\mu_1 \mu_2}]\rho + [a' + \frac{2a\rho'}{\rho}]\mathbf{U}_{scal}. \qquad (2.95)$$

(A word of caution, a in this expression is a function, not a constant.)

Once the basis for the coplanar N-body problem is given below, it is not difficult to show that the same distinction between collinear motion ($c = 0$) and motion where the particles define a straight line ($c \neq 0$) persists. I leave it as an exercise to the reader to examine this and other settings where $\mathbf{W}_{config,rot} \equiv \mathbf{0}$.

2.5.5   Finding central and other configurations

So far we have talked a lot about central configurations, but we have not determined any of them. To avoid resembling that sarcastic definition of a consultant on horses—he knows everything about them, but has never seen one—I now use the above structure to find some central configurations. The idea is natural: use the conditions of Thm. 2.18 to determine when $\nabla_J U$ has no $\mathbf{U}_{scal}$ and $\mathbf{U}_{rot}$ terms. In this way, finding all three-body non-collinear central configurations is easy. While finding the collinear ones can be done in the same way, the conclusion is not immediate, so the discussion is deferred to the next chapter—where an immediate answer is forthcoming.

Theorem 2.19 *For the three-body problem and for any choice of the masses, the only non-collinear central configuration is an equilateral triangle.*

---

[18]Recall (page 73) that if $\mathbf{W}_{config,rot} = \mathbf{0}$, then $\rho_\lambda^2 = c$.

Proof: For a central configuration, $\nabla_J U = -\frac{U}{\rho^2}\boldsymbol{\rho}$, so the α and β values from Thm. 2.18 must be zero. Thus, for a non-collinear configuration, a zero β value occurs iff the configuration is an isosceles triangle where $r_{1,3} = r_{2,3}$. The condition to avoid $\mathbf{U}_{scal}$ terms now reduces to $\rho_1 = r_{1,3} = r_{2,3}$, or an equilateral triangle. $\square$

As an illustration of other ways to use the information concerning the structure of $\nabla_J U$ and the basis for the configurational velocity, we could wonder about the behavior of a solution where, say, $\nabla_J U$ has no $\mathbf{U}_{rot}$ components. What can we say about the motion?

Theorem 2.20 *A non-collinear solution for the coplanar three-body problem never has a $\mathbf{U}_{rot}$ terms for $\nabla_J U$ for a particular choice of Jacobi coordinates iff the motion defines an isosceles triangle for all time. If such motion does not always define an equilateral triangle, then* $\mathbf{c} = \mathbf{0}$.

Proof: The first part is easy; according to Thm. 2.18, such a solution has a $\nabla_J U$ with no $\mathbf{U}_{rot}$ components iff $\beta \equiv 0$ iff the configuration always is an isosceles triangle with $r_{1,3} = r_{2,3}$. Of course, this choice of $\mathbf{U}_{rot}$ requires the particular choice $\boldsymbol{\rho}_1 = \mathbf{r}_2 - \mathbf{r}_1$.

To preserve an isosceles shape, it must be that $\mathbf{W}_{config,rot} \equiv \mathbf{0}$, which means that $b \equiv 0$. Recall from Thm. 2.15 that with an isosceles triangle, $\rho_\lambda = c$. Combining this fact with $\beta \equiv 0$, we have from Eq. 2.94 that $2ac \equiv 0$. If $c \neq 0$, then it must be that $a \equiv 0$ and (from Thm. 2.19) that the configuration always is an equilateral triangle. As the hypothesis excludes this possibility, it must be that $c = 0$. $\square$

Incidentally, Thm. 2.20 does *not* hold for the general three-body problem. Instead, start where $m_1 = m_2$ define an isosceles triangle, with $r_{1,3} = r_{2,3}$, in the x-z plane that is symmetric with respect to the the z-axis and m_3 is on the z-axis. For initial conditions $\mathbf{W}_{rot} \neq \mathbf{0}$, so $\mathbf{c} \neq 0$, and $\mathbf{W}_{config} = \mathbf{0}$, the solution preserves the shape of a (changing) isosceles triangle that rotates about the z-axis. The existence of this motion again indicates that differences arise by replacing $\mathbf{e}_3$ in $\mathbf{U}_{rot}$ with a different and variable normal vector. A discussion of this is deferred to Saari [90].

2.5.6   Equations of motion for constant I

To further illustrate the added information we receive from Thm. 2.18, I now derive the equations of motion when ρ is a constant: they will be in terms of a', b'. It follows from Eq. 2.94 that $\alpha = a' - \frac{b(c+\rho_\lambda^2)}{\rho^2}$ and $\beta = b' + \frac{a(c+\rho_\lambda^2)}{\rho^2}$ while

$\frac{c^2+(\rho \mathbf{W}_{config})^2}{\rho^4} = \frac{U}{\rho^2} = \frac{2|h|}{\rho^2}$ so $\rho^2 = \frac{c^2}{2|h|} + \frac{a^2+b^2}{2\mu_1 \mu_2 |h|}$. The missing equations of motion now reduce to finding a' and b': they are

$$
\begin{aligned}
a' &= \frac{b(c+\rho_\lambda^2)}{\rho^2} + \frac{m_1 \mu_2}{\rho^2}[\rho_1 \rho_2(\frac{m_1}{r_{1,2}^3} + \frac{m_2}{r_{2,3}^3} - \frac{m_1+m_2}{\rho_1^3}) \\
&\qquad + m_3 \cos(\theta)(\frac{1}{r_{1,3}^3} - \frac{1}{r_{2,3}^3})(\frac{\mu_1}{\mu_2}\rho_1^2 - \rho_2^2)] \qquad (2.96) \\
b' &= -\frac{a(c+\rho_\lambda^2)}{\rho^2} - \frac{\mu_1}{\rho_1 \rho_2}[m_3 \mu_1(\frac{1}{r_{1,3}^3} - \frac{1}{r_{2,3}^3})[(\mathbf{e}_3 \times \boldsymbol{\rho}_1) \cdot \boldsymbol{\rho}_2]]
\end{aligned}
$$

The $[\nabla_J U, \mathbf{W}_{config}] = 0$ condition reduces to $aa'+bb' = 0$, or that a^2+b^2 is a constant. Just the appearance of this $a^2 + b^2$ constraint leads one to speculate that if such an orbit could exist, maybe it would be on some sort of ellipse. OK, so this is a huge stretch grasping on what might be purely coincidental connections. But, as described in Sect. 3.2.2 on page 97, this is what would happen.

For completeness, while the equations of motion in Eq. 2.96 are based on the structure of $\nabla_J U$, the representation using Eq. 2.85 is

$$
a' = \frac{\mu_1 \mu_2(\rho_1'' \rho_2 - \rho_2'' \rho_1)}{\rho^2}, \quad b' = \frac{\mu_1 \mu_2[\rho_1 \rho_2(\lambda_1 - \lambda_2)]'}{\rho^2}.
$$

2.5.7   Basis for the coplanar N-body problem

To conclude this chapter, a basis for $\nabla_J U$ and the configurational velocity is given for the coplanar N-body problem. This description is, again, in terms of Jacobi coordinates. Rather than introduce all of the terms, I just move to the conclusions.

For $j = 1, \dots, N-1$, let $\boldsymbol{\rho}_j$ be the position of $\mathbf{1}_{j+1}$ relative to $\frac{m_1 \mathbf{r}_1 + \dots m_j \mathbf{r}_j}{m_1 + \dots + m_j}$, which is the center of mass of the first j particles. Let the generalized masses be defined as $\mu_j = \frac{m_{j+1}(m_1 + \dots + m_j)}{m_1 + \dots + m_{j+1}}$.

With the system inner product $[\mathbf{a}, \mathbf{b}] = \sum_{j=1}^{N-1} \mu_j \mathbf{a}_j \cdot \mathbf{b}_j$, we have that

$$
\boldsymbol{\rho}'' = \nabla_J U;
$$

the usual relationships hold for $\mathbf{V}, \mathbf{W}_{rot}, \mathbf{W}_{scal}$. The following provides the basis to represent the configurational velocity and, by including $\boldsymbol{\rho}, \nabla_J U$.

Theorem 2.21 *For the coplanar N-body problem expressed with Jacobi coordinates, let, for $j = 1, \dots, N-1$, let*

$$
\mathbf{U}_{scal,j} = (\frac{\boldsymbol{\rho}_1}{\mu_1}, \dots, \frac{\boldsymbol{\rho}_j}{\mu_j}, -\frac{\sum_{k=1}^{j} \rho_k^2}{\mu_{j+1} \rho_{j+1}^2} \boldsymbol{\rho}_{j+1}, \mathbf{0}, \dots, \mathbf{0}) \in (\mathbb{R}^3)^{N-1}
$$

and

$$\mathbf{U}_{rot,j} = \mathbf{E}_3 \times \mathbf{U}_{scal,j}.$$

The set $\{\mathbf{U}_{scal,j}, \mathbf{U}_{rot,j}\}_{j=1}^{N-1}$ *is an orthogonal basis for the space of configurational velocity.*

Notice that the vectors $\mathbf{U}_{scal}$ and $\mathbf{U}_{rot}$, introduced earlier in this section for the three-body problem, are $\mathbf{U}_{scal} = \frac{\rho_2}{\rho_1}\mathbf{U}_{scal,1}$ and $\mathbf{U}_{rot} = \frac{\rho_2}{\rho_1}\mathbf{U}_{rot,1}$.

Proof: The different $\mathbf{U}$ terms are non-zero and pairwise orthogonal, so they span a space of dimension $2(N-1)$. The remainder of the proof involves simple computations verifying that $[\boldsymbol{\rho}, \mathbf{U}_{rot,j}] = [\boldsymbol{\rho}, \mathbf{U}_{scal,j}] = [\mathbf{E}_3 \times \boldsymbol{\rho}, \mathbf{U}_{scal,j}] = [\mathbf{E}_3 \times \boldsymbol{\rho}, \mathbf{U}_{rot,j}] = 0.$ $\square$

We have, then, that

$$\mathbf{W}_{config} = \sum_{j=1}^{N-1} [a_j \mathbf{U}_{scal,j} + b_j \mathbf{U}_{rot,j}]$$

and

$$\nabla_J U = -\frac{U}{\rho^2}\boldsymbol{\rho} + \sum_{j=1}^{N-1} [\alpha_j \mathbf{U}_{scal,j} + \beta_j \mathbf{U}_{rot,j}]$$

where

$$a_j = \frac{[\boldsymbol{\rho}', \mathbf{U}_{scal,j}]}{\mathbf{U}_{scal,j}^2}, \quad b_j = \frac{[\boldsymbol{\rho}', \mathbf{U}_{rot,j}]}{\mathbf{U}_{rot,j}^2}$$

and

$$\alpha_j = \frac{[\mathbf{U}_{scal,j}, \nabla_J U]}{\mathbf{U}_{scal,j}^2}, \quad \beta_j = \frac{[\mathbf{U}_{rot,j}, \nabla_J U]}{\mathbf{U}_{rot,j}^2}.$$

Results for the N-body problem that are similar to the above results for the three-body problem now can be obtained.

Chapter 3

Finding Central Configurations

Now that we have discussed the role and importance of central configurations, it is time to find some of them. So far we have only established that the equilateral triangle is a central configuration for the three-body problem (Thm. 2.19 on page 79), but we will find many others. Most approaches introduced in this chapter (particularly starting in Sect. 3.3) differ significantly from what normally is done.

These questions about N-body central configurations have proved to be surprisingly difficult to answer, so it is reasonable to seek other ways to solve some of the problems as well as to develop intuition about these concerns. A natural strategy is to follow the standard mathematical practice of embedding a difficult problem into a larger, more general class of mathematical issues. Whatever the approach, whatever is done to analyze central configurations, part of the goal should be to temper the curse of dimensionality.

To explain the source of this curse, notice that $N = 2$ bodies have only $\binom{2}{2} = 1$ mutual distance. But $N = 3$ bodies define $\binom{3}{2} = 3$ mutual distances. With four and five bodies, the number of mutual distances radically jumps, respectively, to $\binom{4}{2} = 6$ and $\binom{5}{2} = 10$. In other words, after understanding what happens for the three body problem with its three variables, instead of being able to explore what might happen with four, or five variables, the number of mutual distances jumps to six and continues to grow quadratically with N. It would be useful to find an interesting class of problems that involves each number of variables and includes central configurations as a special case.

3.1   From the ancient Greeks to ...

The approach suggested next[1] subsumes the central configuration question into a more general problem where, indeed, it makes perfect sense to analyze what happens with any number of variables and the results are of independent interest. This class of problems traces its origin to issues about geometric figures that were studied by the ancient Greeks. What helps for our purposes is that the status of this topic, as nicely developed by nineteenth century mathematicians such as Cauchy, provides what we need for an initial discussion. An excellent reference for this introduction is the classic book *Inequalities* by Hardy, Littlewood, and Pólya [23].

To motivate what follows, consider the standard calculus problem where the student must find the rectangle with area equal to one that has the smallest perimeter. Can this problem be solved without calculus?

To do so, let x represent the length of one rectangle edge so the other is $y = \frac{1}{x}$ and the perimeter is $2(x + y)$. The problem, then, is to find the minimum value of a sum of positive numbers whose product is unity. For two variables, the answer follows directly from the algebraic expression

$$xy = (\frac{x+y}{2})^2 - (\frac{x-y}{2})^2, \tag{3.1}$$

which leads to the inequality

$$(xy)^{\frac{1}{2}} \leq \frac{1}{2}(x + y) \tag{3.2}$$

where equality holds iff $x = y$. The unique answer to the rectangle problem, then, is $x = y$ or a square.

3.1.1   Arithmetic and geometric means

Equation 3.2 is a special case of the relationship between the arithmetic and geometric means of numbers. To define the terms, if $\mathbf{a} = (a_1, \ldots, a_K)$ represents K positive variables, the arithmetic and geometric means are, respectively,

$$\mathcal{A}(\mathbf{a}) = \frac{1}{K} \sum_{j=1}^{K} a_j \text{ and } \mathcal{G}(\mathbf{a}) = (a_1 a_2 \ldots a_K)^{1/K}.$$

---

[1]Most of the "central configuration" material in this chapter comes from my notes for talks given in the late 1970's at Oberwolfach and elsewhere.

Theorem 3.1 *For $K \geq 1$, the inequality*

$$\mathcal{A}(\mathbf{a}) \geq \mathcal{G}(\mathbf{a}) \tag{3.3}$$

is satisfied where equality holds iff $a_1 = a_2 = \cdots = a_K$.

The following proof of this interesting theorem, which was developed by A. Cauchy in 1821, is included because its transparency and elegance.

Proof: The argument starts with an induction on $K = 2^n$. Because Eq. 3.2 establishes that Eq. 3.3 is true for $K = 2^1$, assume that Eq. 3.3 holds for all powers of two up to $K = 2^n$. Using the induction hypothesis to verify Eq. 3.3 for $K = 2^{n+1}$ we have, by dividing the 2^{n+1} terms into two sets of 2^n terms, that

$$[\Pi_1^{2^{n+1}} a_j]^{1/2^{n+1}} = [(\Pi_1^{2^n} a_j)^{1/2^n} (\Pi_{2^n+1}^{2^{n+1}} a_j)^{1/2^n}]^{1/2}$$
$$\leq [(\tfrac{1}{2^n} \textstyle\sum_1^{2^n} a_j)(\tfrac{1}{2^n} \sum_{2^n+1}^{2^{n+1}} a_j)]^{1/2}.$$

According to Eq. 3.2, the last product is bounded above by the desired

$$\frac{1}{2}[\frac{1}{2^n} \sum_1^{2^n} a_j + \frac{1}{2^n} \sum_{2^n+1}^{2^{n+1}} a_j] = \frac{1}{2^{n+1}} \sum_{j=1}^{2^{n+1}} a_j.$$

The condition for equality, of course, is the equality of all a_j.

It remains to prove the result for any value of $K < 2^m$. The trick here is to add some innocuous terms so that we have 2^m of them. Each new term is what we want at the end—the average of the first K terms. Namely, let

$$b_j = \begin{cases} a_j & \text{for } j = 1, \ldots, K \\ \overline{A} - \frac{1}{K} \sum_{j=1}^K a_j & \text{for } j = K+1, \ldots 2^m \end{cases}$$

Because the conclusion holds for 2^m terms, it follows that

$$[\Pi_{j=1}^{2^m} b_j]^{\frac{1}{2^m}} = \overline{A}^{\frac{(2^m - K)}{2^m}} [\Pi_{j=1}^K a_j]^{\frac{1}{2^m}} \leq \frac{1}{2^m} \sum_{j-1}^{2^m} b_j = \frac{K\overline{A} + (2^m - K)\overline{A}}{2^m} = \overline{A},$$

or $[\Pi_{j=1}^K a_j]\overline{A}^{(2^m - K)} \leq \overline{A}^{2^m}$ and the conclusion $[\Pi_{j-1}^K a_j]^{1/K} \leq \overline{A}$ holds for any integer $K > 0$. Equality occurs iff the a_j all agree. $\square$

Simple illustrations of Thm. 3.1 for $K = 3$ are that the rectangular box with a fixed volume that has the smallest sum of edge lengths is a cube (let the a_j be the edge lengths), and the one with the smallest surface area also is a cube (let the a_j be the surface areas of the different faces). For any K, the same assertion about a rectangular object holds for edges or faces of any dimension.

Weighted means

Our purposes require a generalization of Thm 3.1 that uses other means and weights given by specified positive numbers $(g_1, \ldots, g_K)$.

Definition 3.1 *For $p \neq 0$, the weighted p^{th} mean of the variables $\mathbf{a}$ is*

$$\mathcal{W}_p(\mathbf{a}) = [\frac{\sum_j g_j a_j^p}{\sum_j g_j}]^{1/p}. \tag{3.4}$$

For $p = 0$, the weighted geometric mean[2] is

$$\mathcal{W}_0(\mathbf{a}) = (a_1^{g_1} a_2^{g_2} \ldots a_K^{g_K})^{1/\sum g_j}.$$

The following theorem has been known at least since the 1850s.

Theorem 3.2 *For $p_1 > p_2$,*

$$\mathcal{W}_{p_2}(\mathbf{a}) \leq \mathcal{W}_{p_1}(\mathbf{a}) \tag{3.5}$$

where equality occurs iff $a_1 = a_2 = \cdots = a_K > 0$. Indeed, the sole critical point of $\mathcal{W}_{p_1}(\mathbf{a})/\mathcal{W}_{p_2}(\mathbf{a})$, which is a global minimum, occurs on the diagonal $\mathbf{a} = \alpha(1, 1, \ldots, 1)$, $\alpha > 0$.

A bit later, I will show that the problem of finding central configurations and their properties is a special case of finding critical points of $\mathcal{W}_2(\mathbf{a})/\mathcal{W}_{-1}(\mathbf{a})$ subject to constraints that are imposed on the $\mathbf{a}$ variables. This comment motivates the class of problems that I propose: a class where the issues make sense for each dimension and where finding central configurations becomes a special case. To explain, as Thm. 3.2 identifies the unconstrained critical points of $\mathcal{W}_{p_1}(\mathbf{a})/\mathcal{W}_{p_2}(\mathbf{a})$, a natural extension—and this is the proposed class of problems—is to characterize the critical points when constraints are imposed on the $\mathbf{a}$ variables.

> *Characterize the critical points of $\mathcal{W}_{p_1}(\mathbf{a})/\mathcal{W}_{p_2}(\mathbf{a})$ when the $\mathbf{a}$ variables are restricted to a lower dimensional algebraic set. More generally, find all critical points for a specified algebraic stratification of a portion of $\mathbb{R}_+^K$ that includes the main diagonal.*

Here, $\mathbb{R}_+^K = \{\mathbf{a} = (a_1, \ldots, a_k) \,|\, a_j > 0, j = 1, \ldots, K\}$; it is the interior of the positive orthant of $\mathbb{R}^K$.

---

[2]By using the footnote on page 41 and the description given later in Sect. 3.1.2, it can be shown that the geometric mean is related to vortex problems.

A simple case–a hyperplane

Start with a simple choice: *find the* $\mathcal{W}_{p_1}(\mathbf{a})/\mathcal{W}_{p_2}(\mathbf{a})$ *critical points where* $\mathbf{a}$ *is restricted to a hyperplane passing through the origin.* While the result holds for $p_1 \neq p_2$, restricting attention to $p_1 > 1, p_2 < 0$, permits a trivial convexity-concavity proof.

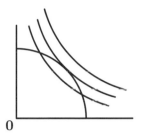

Fig. 3.1. Convex constraint; concave level sets

The argument, as illustrated in Fig. 3.1, is simple. First suppose we are to find the critical points of a strictly concave function relative to a plane: clearly, it is at the unique minimum point. In terms of level sets, this point is where one level set kisses the plane in a unique point: neighboring level sets either miss the plane or intersect it in more than one point. An even easier problem is to replace the plane with a convex level set as it folds in the opposite direction: this choice corresponds to the posed problem.

We have the obvious, easily proved fact that restricting a strictly concave function to a level set of a strictly convex function has only one critical point: it is where the concave function has a minimum value. Again, the reason is that at any other intersection point, the level sets have a transverse intersection preventing the gradient of the concave function from being in the normal bundle of the convex function.[3] The same argument holds even if the level set of the strictly convex function is intersected with a lower dimensional hyperplane that passes through $\mathbf{0}$.

Theorem 3.3 *Let* $\mathbf{a} = (a_1, \ldots, a_K)$ *be restricted to a k-dimensional hyperplane* $\mathcal{P}$ *in* $\mathbb{R}^K$, $K \geq 2$ *and* $1 \leq k < K$, *that passes through the origin and* $\mathbb{R}^K_+$. *For* $p_1 > 1$ *and* $p_2 < 0$, *the only critical points of* $\mathcal{W}_{p_1}(\mathbf{a})/\mathcal{W}_{p_2}(\mathbf{a})$ *in* $\mathcal{P} \cap \mathbb{R}^K_+$ *are on one line* $t(a_1^*, \ldots, a_K^*)$.

Proof: If the positive $a_1 = a_2 = \cdots = a_K > 0$ diagonal is in the hyperplane, it is the line. So assume that this diagonal is not in the hyperplane.

---

[3]The same transverse intersection comment holds when p_1 and p_2 have the same sign, but details about the curvature need to be added. Indeed, this argument can be used to prove the classic Thm. 3.2.

If $\mathbf{a}^* \in \mathcal{P} \cap \mathbb{R}_+^K$, then, since $\mathcal{P}$ passes through the origin, the line $t\mathbf{a}^*$ also is in $\mathcal{P}$. As $\mathcal{W}_{p_1}(\mathbf{a})/\mathcal{W}_{p_2}(\mathbf{a})$ is homogeneous of degree zero, if $\mathbf{a}^* = (a_1^*, \ldots, a_K^*)$ is a critical point, then so are all points on the line $\{t\mathbf{a}^* \,|\, t > 0\}$. This assertion along with the homogeneity of the quotient allows us to restrict attention to $\mathcal{W}_{p_1}(\mathbf{a}) = 1$. Namely, find the critical points of $\mathcal{W}_{p_2}(\mathbf{a})$ when restricted to the manifold defined by $\{\mathbf{a} \in \mathbb{R}_+^K \,|\, \mathcal{W}_{p_1}(\mathbf{a}) = 1\}$ and the plane $\mathcal{P}$—a lower dimensional strictly convex surface.

The conclusion follows from the geometry. For $p_1 > 1$, the level sets of $\mathcal{W}_{p_1}$ are strictly convex: interior points on a line segment connecting two points on a $\mathcal{W}_{p_1}$ level set are between $\mathbf{0}$ and the level set; e.g., $\mathcal{W}_2 = 1$ defines the portion of a sphere in the positive orthant. Consequently, the intersection of the $\mathcal{W}_{p_1} = 1$ level set with $\mathcal{P}$ defines a smooth $(k-1)$-dimensional strictly convex surface.

The $\mathcal{W}_{p_2}$ level sets, $p_2 < 0$, which are the same as the level sets for $\sum_j \frac{g_j}{a_j^{|p_2|}}$, are strictly concave: interior points on a line segment connecting two points on a $\mathcal{W}_{p_2}$ level set are separated from $\mathbf{0}$ by the level set. It trivially follows that there is a unique critical point, so the theorem is proved. $\square$

3.1.2   Connection with central configurations

To connect Thm. 3.2 with the N-body problem, set $K = \binom{N}{2} = \frac{N(N-1)}{2}$, $g_i = m_j m_k$, and $a_i = r_{jk}$. With these choices, $I^{1/2} = A\mathcal{W}_2$ and $U^{-1} = B\mathcal{W}_{-1}$ become, respectively, the quadratic and harmonic means of the mutual distances where A and B are scalar multiples depending on the masses. Thus the critical points of the configurational measure, $\nabla IU^2 = 0$, agree with the critical points of

$$I^{1/2}U = A\frac{\mathcal{W}_2}{\mathcal{W}_1},$$

where A is another positive constant based on the masses. (Throughout, capital letters such as A and B refer to positive constants that most surely change value with each usage.)

Equilateral N-gons

Restating Thm. 3.2 in terms of I and U, it follows for any N that one central configuration is an equilateral N-gon where $r_{jk} = r_{ab}$ for all $j \neq k, a \neq b$. Such a geometric figure resides in a $N-1$ dimensional space, so this conclusion is meaningful only for $N = 3, 4$.

Theorem 3.4 *For any choice of positive masses, for $N = 3$ the only non-collinear central configuration is an equilateral triangle. For $N = 4$, the only non-planar central configuration is an equilateral tetrahedron.*

That these configurations are central configurations follows from the weighted mean argument. That they are unique is developed later.

What adds interest to this theoretical construct is that the equilateral triangle configuration is actually observed in our solar system. To search for this configuration, a natural choice would be to let the two heaviest bodies, Sun and Jupiter, create the defining leg of two equilateral triangles that lie in the plane of motion. Theorem 3.4 suggests examining the vertices of both triangles to see if anything can be found. When astronomers finally got around to doing so, they discovered asteroids, now called the Trojans. But they sure took their time to conduct this search: Lagrange discovered these equilateral triangle configurations in 1772, and 134 years later, in February 1906, the German astronomer Max Wolf (1863-1932) discovered these asteroids by searching the equilateral triangle locations.[4]

Lehmann-Filhés [35] is given credit for discovering the equilateral tetrahedron configuration in 1891—over a century after Lagrange's triangle configuration.[5] The normal way to find these configurations is with direct (but laborious) computations. The normalized weighted mean approach to study central configurations (my preferred approach since I started studying the N body problem) is simpler—finding and verifying the equilateral configurations is an essentially free conclusion—while the analysis suggests the above wider class of closely related mathematical problems.

---

[4]Wolf, from the University of Heidelberg, named the asteroids after Greek heros: the ones that lead Jupiter are named for heros of the Iliad, the trailing ones for heros of Troy. Rather than staying at the triangle vertex, the asteroids move in orbits tracing out a banana that is centered about the vertex—the orbits stay close to what are known as "Hill curves" in the restricted three-body problem. An interesting question is to understand why this happens. Another fascinating mystery is that there are more asteroids in the leading group than in the tailing one: this sounds like a stability issue that can be addressed via dynamics. Incidentally, these Lagrange points occur elsewhere. In 1990, for instance, an asteroid, called Eureka, was found at the triangle point in the orbit of Mars. And, there may be asteroids at the Lagrange points of Venus and even of the Earth. Such behavior may even exist at the triangle points defined by the Earth and our moon—or so Frederic Petit, the director of the observatory of Toulouse, believed in 1846 when he claimed that a second moon of the Earth had been discovered. Sounds silly, but, then, even if this "moon" just consisted of dust with dynamical instabilities that forced everything to fade and disappear in a short time, why not? (Because the mass of the moon is about $\frac{1}{81}$ that of the Earth, it may be possible to have stable "dust" motion.)

[5]I learned this among other interesting historical facts from A. Wintner's classic [112].

Collinear central configurations

The relevance of the constrained optimization question (page 86), which generalizes the search for N-body central configurations, is that for particles in the N-body problem to reside in a certain dimensional physical space, constraints must be imposed on the r_{jk} variables. For instance, it is physically impossible to construct a collinear configuration where $r_{12} = r_{23} = r_{13}$, or to construct any three body configuration where $r_{12} = r_{23} = 1$, $r_{13} = 3$. Instead, the mutual distances among particles must satisfy the triangle inequality, and those for a collinear configuration must satisfy the equality portion corresponding to the degenerate triangle inequality. Thus, for three particles, there are three "collinear" constraints depending on the ordering of the particles; e.g., the constraint when particle 2 is in the middle is

$$r_{12} + r_{23} = r_{13}. \tag{3.6}$$

When described in terms of $x = r_{12}$, $y = r_{23}$, $z = r_{13}$, it becomes clear that the collinear configurations with this ordering reside in the plane $x + y = z$. For $N > 3$, the collinear constraints consist of a collection of linear equalities: in a space of mutual distances, they define a lower dimensional hyperplane. Thus, a natural first choice, already considered in Thm. 3.3, is to find the $\mathcal{W}_{p_1}(\mathbf{a})/\mathcal{W}_{p_2}(\mathbf{a})$ critical points on a plane passing through the origin. Restating Thm. 3.3 in terms of $\mathcal{W}_2(\mathbf{a})/\mathcal{W}_{-1}(\mathbf{a})$, we obtain a much simpler proof (on page 91) of the following classical result.

Theorem 3.5 *(Moulton [51]) For $N \geq 3$, there are precisely $N!/2$ collinear central configurations. In particular, each arrangement of the particles on the line defines precisely one central configuration.*

To indicate some of the history, if particle 2 is in the middle of a three-body configuration, then by setting $a = r_{23}/r_{12}$, the central configuration Eq. 2.17 can be expressed as the root of a quintic polynomial in a. In 1767, Leonhard Euler (1707-1783) developed and analyzed this quintic equation to discover the three collinear configurations for the three-body problem.[6] Moulton [51] found the result for all $N \geq 3$.

It is standard to joke about trying to solve the N-body problem by using induction: Moulton actually did so for this particular problem.[7] He starts by

---

[6]To use a new and different equation, the same approach can be applied to Eq. 2.93 (page 76) to find when $\alpha = 0$.

[7]Forest Ray Moulton (1872-1952) was an astronomer at the University of Chicago. His work in celestial mechanics includes (with Thomas Chamberlin in 1906), the

letting $\mathcal{N}_k$ be the number of collinear central configurations for k particles. Next, he shows that by adding an infinitesimal mass m_{k+1}, the number of solutions becomes $(k+1)\mathcal{N}_k$. This makes sense: the k original particles along a line define $(k+1)$ slots to place the infinitesimal particle so Moulton shows that in each slot, there is a unique position to place the infinitesimal particle to create a central configuration. (The infinitesimal particle does not affect the position of the first k particles, but the other particles most surely determine the position of the infinitesimal one.)

Moulton then shows that "[as] the infinitesimal mass m_{k+1} increases continuously to any finite positive value," the number of central configurations remains $(k+1)\mathcal{N}_k$. In other words, the location of this formally-known-as-the-infinitesimal-particle might change, and it might alter the position of the other particles, but Moulton proves that there are no critical points allowing a bifurcation that would change the number of central configurations. Thus, $\mathcal{N}_{k+1} = (k+1)\mathcal{N}_k$. Because Euler showed that $\mathcal{N}_3 = \frac{3!}{2}$, the conclusion follows.

Incidentally, in this paper, Moulton also addresses the converse problem. Namely, start with any positioning of particles along the line. Can mass values be found that make these positions a central configuration? Because the masses occur linearly in the numerators, this problem is easier to analyze.

The following geometric proof of Moulton's Theorem, which is much simpler and very different in nature, comes from Saari [85].

Simple proof of Thm. 3.5: Each ordering of particles along a line defines a particular plane in the space of mutual distances. According to Thm. 3.3, each such plane defines a unique collinear central configuration. As there are $N!/2$ orderings (up to rotation), the conclusion follows. $\square$

These results describe all collinear central configurations and all central configurations in a $N-1$ dimensional physical space. As such, the story is completed for $N = 3$: the four central configurations, which define five positions of m_3 relative to m_1 and m_2, are indicated in Fig. 3.2.

The next challenge is to determine the coplanar central configurations for $N > 3$ and the non-coplanar central configurations for $N > 4$. Complete

---

"Chamerberlin-Moulton planetesimal hypothesis:" it concerns the origin of the planets in terms of "capture" that might occur should particles move too close to the sun. He was the Permanent Secretary and then Administrative Secretary of the Amer. Assoc. Adv. Science from 1937-48. Most people working in celestial mechanics know his classic *An Introduction to Celestial Mechanics*, which fortunately has been reprinted by Dover [52], but he also made advances in the numerical integration of differential equations and wrote a book on this topic. In 1932 he made the bold prediction that there was no hope of reaching the moon "because of the insurmountable barriers to escaping the earth's gravity."

answers are nowhere near being found. Useful insights, however, should arise from examining the critical points of $\mathcal{W}_{p_1}(\mathbf{a})/\mathcal{W}_{p_2}(\mathbf{a})$ subject to the specified kinds of constraints. The reason we should investigate algebraic and stratified constraints is described next.

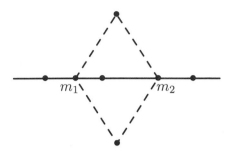

Fig. 3.2. Central configuration positions for m_3

3.2   Constraints

The highly impatient reader could jump immediately to the introduction of a geometric approach to find central configurations (Sect. 3.3, page 102) even though we first should understand the natural constraints that are imposed on this problem. To start the description, let $\mathbf{R} = (\mathbf{r}_1, \ldots, \mathbf{r}_N) \in (\mathbb{R}^q)^N$ be the positions of the N particles in a q-dimensional physical space where, instead of $q = 1, 2$, or 3, initially allow any value up to $q = N - 1$. Let

$$F = (F_{1,2}, \ldots, F_{N-1,N}) : (\mathbb{R}^q)^N \to \mathbb{R}_+^{\binom{N}{2}}$$

be the mapping that describes the mutual distance between particles; i.e., $F_{j,k}(\mathbf{R}) = |\mathbf{r}_j - \mathbf{r}_k|$. This space $\mathbb{R}_+^{\binom{N}{2}}$ is used often enough to deserve a name; so , unimaginatively, call it the "space of mutual distances" or SMD. When the dimension, $\binom{N}{2}$, should be mentioned, I use $SMD(\binom{N}{2})$.

Let the $SMD(\binom{N}{2})$ variables be ξ_{jk}, $j < k$, and define $\tilde{I}, \tilde{U} : \mathbb{R}_+^{\binom{N}{2}} \to \mathbb{R}$ as

$$\tilde{I} = \frac{1}{2M} \sum_{j<k} m_j m_k \xi_{j,k}^2, \quad \tilde{U} = \sum_{j<k} \frac{m_j m_k}{\xi_{j,k}}. \tag{3.7}$$

With these definitions, we have $I = \tilde{I} \circ F(\mathbf{R})$, $U = \tilde{U} \circ F(\mathbf{R})$. The physical constraints needed to analyze central configurations are characterized by the image of $F(\mathbf{R})$. The connection is given by

$$\nabla I U^{1/2}(\mathbf{R}) = \nabla \tilde{I} \tilde{U}^{1/2} \circ D_{\mathbf{R}} F \tag{3.8}$$

where the image of $D_{\mathbf{R}}F$ defines a SMD tangent space anchored at $F(\mathbf{R})$. In other words, $\mathbf{R}^*$ is a central configuration if and only if $\nabla \tilde{I}\tilde{U}^{1/2}$ is in the normal bundle of the $D_{\mathbf{R}^*}F$ tangent space.

A way to appreciate the advantages of this approach is to recognize that it is difficult to find central configurations because of the constraints imposed on which configurations can be constructed in different dimensional physical spaces. To see how these physical constraints complicate any analysis of U, compare the complex appearing

$$\nabla U = (\ldots, \sum_{j=1}^{N} \frac{m_j m_k (\mathbf{r}_k - \mathbf{r}_j)}{r_{j,k}^3}, \ldots) \tag{3.9}$$

with the much simpler

$$\nabla \tilde{U} = (\ldots, -\frac{m_j m_k}{\xi_{j,k}^2}, \ldots). \tag{3.10}$$

The difference between the expressions is captured by the components of

$$\nabla U = \nabla \tilde{U} \circ DF. \tag{3.11}$$

Namely, as displayed by Eq. 3.11, using $U = \tilde{U} \circ F(\mathbf{R})$ separates the central configuration problem into terms describing the physical constraints needed to construct a configuration, as governed by F, from the structures peculiar to the N-body problem, as captured by $\tilde{I}$ and $\tilde{U}$. As Eq. 3.9 demonstrates, the complicated structure of ∇U is caused by the summations that reflect the DF constraints. In turn, the simple structure of the self-potential is exhibited in Eq. 3.10.

The level sets of $\tilde{I}$ and $\tilde{U}$ are simple: one is strictly convex and the other is strictly concave. This approach also identifies the role played by the masses: the $m_j m_k$ values affect the geometry of the level sets of $\tilde{I}$ and $\tilde{U}$, but they play no role in the image of F. Thus, a natural search approach is to determine all central configurations, if any exist, on the different components of the F image. To do so requires examining F's singularity structure.

As an illustration, a standard argument and computation shows that F is an open mapping when it maps configurations spanning a $(N-1)$-dimensional space to $\mathrm{SMD}(\binom{N}{2})$. As F imposes no restrictions, the only $(N-1)$-dimensional configurations are the critical points of $\nabla \tilde{I}\tilde{U}^2$—the equilateral configurations.

3.2.1   Singularity structure of F

To use Eq. 3.8, we need to determine the singularity structure of F; e.g., we need to determine where the rank of DF changes and what configurations **R** are in the kernel of $D_{\mathbf{R}}F$. Certain aspects of the singularity structure of F are immediate. For instance, $F(\mathbf{R})$ is rotation invariant: rotating a configuration **R** as a rigid body does not change any mutual distances, so all of these rotations have the same $F(\mathbf{R})$ image. This means that all rigid body rotations remain in the kernel of $D_{\mathbf{R}}F$. Similarly, if the N-particles defining configuration **R** span a q_1-dimensional space, then all higher dimensional configurations are in the kernel of $D_{\mathbf{R}}F$.

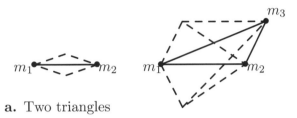

a. Two triangles

b. Two tetrahedrons

Fig. 3.3. Same lengths, two choices = singularity of F

A folding

The reason for the last assertion is suggested by the two triangle configurations of Fig. 3.2 (page 92) or the two dashed triangles in Fig. 3.3. Both have the same F image, so the image must have be "folded"—similar to how $y = x^2$ "folds" the x-axis near $x = 0$. This singularity structure manifests the fact that the reflection of a $(q_1 + 1)$-dimensional configuration about a q_1-dimensional plane defines a second configuration with the same F image. (In a $(q_1 + 1)$-dimensional space, the reflected image is not captured by a rotation.) Rather than carrying out the straightforward (but messy) computation establishing that DF has a lower rank, it is more useful to provide some intuition.

The $f(x) = x^2$ critical point at $x = 0$ occurs because $x = \pm\epsilon$ have the same y value, so the critical point captures the folding of the image about 0—this is a special case of a general phenomenon. After all,

$$0 = \frac{f(\epsilon) - f(-\epsilon)}{2\epsilon} = \frac{1}{2}\left[\frac{f(\epsilon) - f(0)}{\epsilon} + \frac{f(-\epsilon) - f(0)}{-\epsilon}\right] \to f'(0) \text{ as } \epsilon \to 0.$$

Similarly, the leg lengths of a $(q_1 + 1)$-dimensional configuration define a configuration and its reflection with the same F image: their images are "folded" about the space of q_1-dimensional configurations. In Fig. 3.3a, for instance, the dashed lines above and below the solid line have the same lengths and define two different triangles with the same F values. As the minimal altitude of each dashed triangle approaches zero—to define a one-dimensional configuration—the rank of DF drops. A similar description holds with Fig. 3.3b where the two sets of three dashed lines define two tetrahedrons—reflections of each other—with the same F image. Again, a singularity occurs as the dashed lines shrink so that the figure becomes two-dimensional. (Other singularities of F are described later.)

Dimensions of different configurations

Our search for central configurations requires knowing the dimensions of the F image of different physical dimensional configurations; e.g., the collinear, coplanar, or ... configurations.

Theorem 3.6 *For a N-body configuration, the collinear configurations define $(N - 1)$-dimensional linear subspaces in $SMD(\binom{N}{2})$. In general, the N-body configurations that span a q-dimensional subspace in physical space are in $Nq - \binom{q+1}{2}$ dimensional SMD submanifolds or their boundaries.*[8]

A way to develop intuition for the asserted dimensions is to construct a coplanar configuration by starting with three particles and the triangle they define. When a new particle is added at any stage, it defines a triangle with some two particles (vertices) already in the system (so their leg length is determined). Thus, the location of the new particle requires selecting $q = 2$ leg lengths that satisfy the triangle inequality. But notice, the choice of these lengths define *two triangles*: they are reflections about the line connecting the original two. Whichever choice is made, all remaining mutual distances follow from the geometry. So, there are three variables to start and two more for each of the remaining $N - 3$ particles leading to the stated dimension of $3 + 2(N - 3) = 2N - \binom{2+1}{2}$.

Similarly, to construct three-dimensional configurations, start with a tetrahedron that is defined by four particles and six mutual distances. Each new particle creates another tetrahedron where some three previously introduced particles serve as vertices. Since the leg lengths among the original particles are determined, only the $q = 3$ lengths connecting each new particle are free to be selected. This leads to the dimension $6+3(N-4) = 3N-\binom{3+1}{2}$.

---

[8]The "boundary" comment is explained in Sect. 3.2.3.

Proof: Since the F image of collinear configurations lies in SMD hyperplanes (page 90), only the dimensions need to be computed. With a specified ordering, the $N - 1$ mutual distances between adjacent particles defines the configuration: the dimension is $N - 1$.

A q-dimensional configuration starts with $(q + 1)$ particles that span a q-dimensional space of physical space. These mutual distances are free to be selected (subject to constraints such as the triangle inequality), so this introduces $\binom{q+1}{2}$ variables. Adding a new particle defines, along with q of the previously introduced particles, a q-gon. The mutual distances among the q previously selected particles are determined, so only the mutual distances for the legs connecting these q particles with the new one are free to be selected. All other mutual distances are uniquely determined by these values and the geometry.

For a dimension count, there are $\binom{q+1}{2}$ distances to be selected for the first $q + 1$ particles. Each of the remaining $N - (q + 1)$ particles adds q dimensions for $Nq - (q + 1)q + \binom{q+1}{2} = Nq - \binom{q+1}{2}$ variables (mutual distances) to be selected (up to geometric constraints). The geometry of physical space determines the remaining values. $\square$

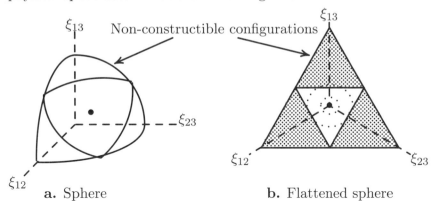

Fig. 3.4. Three-body constraints

Three-body construction

The three-body situation is depicted in Fig. 3.4a (also see Easton [19]) where the portion of the sphere is a level set of $\tilde{I}$. (Figure 3.4b is a flattened version of the sphere.) The three lines depict the intersection of the sphere with the three planes of collinear configurations: they differ by which particle is in the middle. These planes (or lines on the sphere) must intersect along a $\xi_{j,k} = 0$ coordinate plane: this intersection corresponds to what would be a collision between particles m_j and m_k.

The interior of the region is where the ξ_{ij} variables satisfy the triangle inequality so they represent configurations that can be physically constructed—each point defines two configurations that differ by reflection. For instance, the center point (denoted by a bullet) where $\xi_{1,2} = \xi_{1,3} = \xi_{2,3}$ represents two equilateral triangles: they are distinguished by whether particle two is adjacent to particle one in a clockwise, or counterclockwise, direction. Points in the remaining three regions—the three shaded regions in the Fig. 3.4b flattened sphere—fail to satisfy the triangle inequality, so they do not correspond to physically constructible configurations.

What about the two elliptical dotted arcs in Fig. 3.4? I explain them next.

3.2.2 Some dynamics

The next natural step is to continue describing the singularity structure of F, but let's take a short break from dimension-counting to partially relate Fig. 3.4 to N-body dynamics and, in particular, to the velocity decomposition introduced in Chap. 2. (This material is not needed to understand the rest of the chapter.)

Level sets of $\dot{U}$

For an illustrating issue, return to the earlier conjecture (page 47) that a constant I (or R) requires a relative equilibria motion. In Sect. 2.3, it is shown that if I is a constant, say D, then $U = 2|h|$ where h is the total energy. Thus if the conjecture is false, then there is some orbit, which is not a relative equilibrium orbit, where $I = D$ and $U = 2|h|$. What can we say about this orbit? The complicated form of U suggests the natural first guess, "Very little."

Some people had interesting ideas. During informal discussions at a 2003 conference at the University of Maryland on the N-body problem, some participants were speculating about how to find a counter-example for the conjecture. A candidate at the time was a particular three-body orbit that passes through the equilateral triangle and then through the collinear central configurations: could it have a constant I value? It cannot.

To prove this, use the above structure of $\tilde{I}, \tilde{U}$ and Fig. 3.4. After all, any orbit satisfying $I = D$ and $U = 2|h|$ is mapped by $F(\mathbf{R})$ to the set of points where $\tilde{I} = D$ and $\tilde{U} = 2|h|$. While the level set of $U - 2|h|$ may be complicated, the level set of $\tilde{U} = 2|h|$ is simple: in Fig. 3.4, it is a closed curve on the $\tilde{I} = D$ surface that roughly resembles an ellipses enclosing the center point $\xi_{1,2} = \xi_{1,3} = \xi_{2,3}$ (corresponding to an equilateral triangle). The two dotted ellipses in Fig. 3.4b depict two $\tilde{U}$ level sets.

So, motion satisfying $I = D$ and $U = 2|h|$ must have orbits resembling an ellipse such as the dotted ones in Fig. 3.4b: this confirms the earlier speculated comment on page 81. The regularity of the level sets allows us to immediately dismiss the conjecture about such an orbit passing through the equilateral triangle configuration. For this to occur, the level set of $\tilde{U}$ would have to pass through the point $\xi_{1,2} = \xi_{1,3} = \xi_{2,3}$. But, as we learned from the proof of Thm. 3.4, this particular $\tilde{U}$ level set is just that one point. So, if a constant I orbit passes through the equilateral triangle, the motion preserves this configuration for all time: the conjecture holds in this setting.[9]

To leave the reader with a challenge, can a constant I orbit have both collinear and non-collinear configurations? To suggest a way to analyze this question, notice that, *if the $\tilde{U}$ level set does not define a collinear central configuration,* it transversely intersects an edge of the Fig. 3.4b triangle. Thus any motion constrained to this level set that creates a collinear configuration must make a sharp turn. After all, the motion cannot continue going in the same direction along the $\tilde{U}$ level set as it would then enter a region of impossible configurations. Consequently, the motion must make an abrupt 360° change to retrace its path on the $\tilde{U}$ level set. This abrupt change should cause a "singularity," maybe in $\mathbf{W}_{config}$.

Velocity decomposition terms

To answer the kind of challenge just issued, it helps to understand how to represent the various velocity components

$$\mathbf{V} = \mathbf{W}_{scal} + \mathbf{W}_{rot} + (\mathbf{W}_{config,scal} + \mathbf{W}_{config,rot}),$$

in a figure such as Fig. 3.4. (See Thm. 2.3 and the discussion of Sect. 2.3.)

Because $F(\mathbf{R}) : (\mathbb{R}^3)^N \to \mathbb{R}_+^{\binom{N}{2}}$, DF maps tangent vectors of $(\mathbb{R}^3)^N$ to tangent vectors of $\mathbb{R}_+^{\binom{N}{2}}$. That is, DF maps $\mathbf{V}$ to a SMD tangent space.

To identify the $DF(\mathbf{V})$ components, the rotation invariance of $F(\mathbf{R})$ requires $DF(\mathbf{W}_{rot}) = \mathbf{0}$: the system rotational velocity is in the kernel of DF. As $\mathbf{W}_{scal}$ changes the size of R, or I, $DF(\mathbf{W}_{scal})$ is a vector orthogonal to the portion of the sphere in Fig. 3.4a representing a $\tilde{I}$ level set.

---

[9]A difference between the Chap. 2 discussion about a fixed R and a fixed configurational measure RU is that a constant RU orbit is on the cone defined by the SMD origin and a $\tilde{U}$ level set for a fixed $\tilde{I}$.

The system configurational velocity, which alters the shape of the configuration, remains. Clearly, $\mathbf{W}_{config}$ plays an important role in the SMD discussion. But, as indicated in earlier discussions about $\mathbf{W}_{config,rot}$ (e.g., page 71), a complication arises because the description depends on the choice of the Jacobi coordinates. To delay addressing these complications, start with $DF(\mathbf{W}_{config,scal})$ when the configuration is collinear. Here, $F(\mathbf{R})$ is a point on an edge of the interior triangle of Fig. 3.4b. As $\mathbf{W}_{config,scal}$ changes the relative ρ_1, ρ_2 distances, but not the angle between ρ_1, ρ_2, $DF(\mathbf{W}_{config,scal})$ is along that edge of the triangle.

Compare this description with what happens when $\mathbf{R}$ defines a triangle where $F(\mathbf{R})$ is point in the interior of the Fig. 3.4b unshaded triangle. This velocity term keeps fixed the angle between ρ_1, ρ_2, but changes the relative sizes of the lengths of the two vectors—one grows and the other shrinks—so different r_{jk} values change. Thus $DF(\mathbf{W}_{config,scal})$ is *not* parallel to the edges of the interior triangle of Fig. 3.4b.

To discuss $\mathbf{W}_{config,rot}$ suppose, as in Thm. 2.14, that $\rho_1 = \mathbf{r}_2 - \mathbf{r}_1$. Now, $\mathbf{W}_{config,rot}$ does not change the lengths of ρ_1, ρ_2, but it does change the angle between them. While this change in angle does not affect $r_{1,2}$, it does change the $r_{1,3}, r_{2,3}$ values. Consequently, $DF(\mathbf{W}_{config,rot})$ is parallel to a line of fixed $\xi_{1,2}$ values: in Fig. 3.4b, it would be parallel to the right-hand edge of the triangle. Other choices for Jacobi coordinates have $DF(\mathbf{W}_{config,rot})$ parallel to other edges of the triangle.

An interesting feature about the singularity structure of F occurs whenever a three-body motion passes through a collinear configuration. When $\mathbf{R}$ is a collinear configuration, the higher dimensional configurations are in the kernel of DF. But all terms in $\mathbf{U}_{rot}$ (Thm. 2.14, page 71) are orthogonal to the line, so we have that $DF(\mathbf{W}_{config,rot}) = \mathbf{0}$.

3.2.3 Stratified structure of the image of F

After this slight digression, it is time to return to the singularity structure of F. As we must anticipate, F defines a stratified structure. To remind the reader what this means with a simpler mapping $f(x,y) = (xy, \frac{1}{2}(x^2 + y^2))$ where

$$Df = \begin{pmatrix} y & x \\ x & y \end{pmatrix},$$

notice that Df has rank zero at point $(x,y) = (0,0)$. The matrix Df has rank 1 iff $(x,y) = a(y,x)$ for some scalar $a \neq 0$. That is, it has rank 1 on the four line segments $y = x, x \neq 0$, and $y = -x, x \neq 0$. Finally, Df has rank 2 everywhere on $\mathbb{R}^2$ off of the lines $y = x, y = -x$.

Notice the "stratified structure:"

- the region where Df has rank zero is the boundary for the region where Df has rank one;

- the region where Df has rank one forms the boundary for the region where Df has rank two.

The same feature holds for F: the rank of DF changes where configurations of one type define the boundary of configurations of another type.

Three-body structure

To describe this stratified effect for F with $N = 3$, the image of F is the closed cone in Fig. 3.4a (or, the non-shaded portion of Fig. 3.4b) defined by the three three planes (represented as lines on a $\tilde{I} = 1$ surface). The image of F has a stratified structure where the interior of the cone represents the two-dimensional configurations. The boundary of this three-dimensional cone consists of the three two-dimensional planes: they correspond to the collinear configurations. In turn, the boundaries of each plane are where the plane intersects a $\mathbb{R}^3$ coordinate plane: this one dimensional line represents a collision. Each of these one-dimensional collision lines has the origin as a boundary: the origin corresponds to a complete collapse of the system.

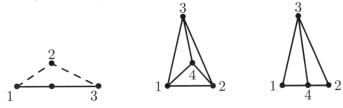

a. Degenerate   **b.** Degenerate   **c.** Really
     Triangle   Tetrahedron   Degenerate!

Fig. 3.5. Degencrates creating boundary conditions

To explain this stratified behavior, consider Fig. 3.5a where the triangle defined by the dashed lines represents a general, two-dimensional configuration. As one vertex approaches the line defined by the other two—the configuration approaches a degenerate triangle—we have a collinear configuration. Namely (reflected by the continuity of F and Fig. 3.5a), the boundary of the three-body coplanar configurations are the collinear configurations. Moreover, a natural constraint for the $\xi_{i,j}$ variables to represent a collinear configuration are those characterizing a degenerate triangle: triplets of the $\xi_{i,j}$'s define equality for the triangle inequality.

Four-body structure

With interesting differences, a similar stratification holds for $N = 4$. The basic three-dimensional four-body figure is a tetrahedron, so the boundary configuration arises when one vertex passes through the plane defined by the three other vertices: this means that the boundary of the three-dimensional configurations are the two-dimensional ones. Thus, constraints describing coplanar configurations are conditions identifying degenerate tetrahedrons as depicted in Fig. 3.5b; e.g., a choice currently is in favor and introduced in 1900 by Dziobek [18] is to use the algebraic determinant conditions specifying a zero volume for the tetrahedron. As shown later, there are other choices of constraints that seem easier to use.

The new feature is depicted in Fig. 3.5c. In the degenerate tetrahedron of Fig. 3.5b, five of the six leg lengths must be specified to determine the configuration—these configurations lie in five-dimensional subspaces of the SMD($\binom{4}{2}$). But Fig. 3.5c requires only *four* distances: knowing the three leg lengths of the large triangle and the $r_{2,4}$ distance suffices to determine all six mutual distances. Thus configurations of the Fig. 3.5c type are in a four, not a five, $SMD(6)$ dimensional subspace. To suggest why $D_\mathbf{R}F$ drops a rank, recall (page 95) that the leg lengths of a new particle define two triangles that are reflections about a line. The Fig. 3.5c configuration is where particle 4 is at the critical point, or a singularity of F.

A geometric description of the $N = 4$ stratified structure of the F-image of SMD(6) now is immediate.

- The collinear configurations define 12 three-dimensional submanifolds in the F image in SMD.

- The collinear configurations form the boundaries for four-dimensional submanifolds in the F image; this larger space corresponds to configurations where just one particle is moved off of the line to create a version of Fig. 3.5c. Each choice of "leg lengths" for this particle off of the line defines two configurations: they are reflections about the line.

- These Fig. 3.5c configurations generate SMD four-dimensional manifolds that form the boundary for five-dimensional portions of SMD. This larger portion comes from general two-dimensional configurations generated by moving another particle off of the original line.

- The image of the coplanar configurations form the boundaries of the non-coplanar three-dimensional configurations generated by lifting one particle from a Fig. 3.5b configuration off the plane.

- The image of three-dimensional configurations form an open set that includes the point corresponding to an equilateral tetrahedron.

N-body

The description for any N is similar. The open region of the F image is defined with $(N-1)$-gons, and its boundaries are defined by degenerate $(N-1)$-gons where one vertex is in the space defined by $N-1$ of the other vertices: the process continues. Alternatively, start with particles on the line that defines $N!/2$ different $(N-1)$-dimensional linear spaces. This space cannot be the full boundary for the general coplanar configurations as they have dimension $2N-3$: instead it is the boundary for those configurations created when one particle is lifted off the line. These configurations define the boundary for the configurations caused by lifting two particles from the line. Continue until $N-2$ particles have been lifted where no three are on a line segment.[10] The two dimensional configurations are the boundary for the three dimensional configurations and they are created by lifting one particle at a time off of the plane. Here, configurations corresponding to Fig. 3.5c have four particles in a plane.

To connect this discussion with the general problem that is posed on page 86 about finding critical points of $\mathcal{W}_{p_1}/\mathcal{W}_{p_2}$ subject to constraints, notice that finding central configurations involves using the physical conditions imposed on the mutual distances required to be able to construct a configuration; i.e., the stratified structure of the image of F. As these constraints generate an algebraic stratification for a portion of the SMD, the posed problem includes central configurations as a special case.

3.3 Geometric approach—the rule of signs

The construction of the last section provides an easy way to find certain properties of central configurations: the approach is called the "rule of signs." First replace the stratified image of F with geometric constraints $g_j(\xi) = 0$. For the three-body problem, for instance, $g(\xi) = \xi_{1,2} + \xi_{2,3} - \xi_{1,3} = 0$ defines a collinear constraint for the ordering where particle 2 is in the middle. In this manner, instead of using the tangent plane of $D_{\mathbf{R}^*}F$ in Eq. 3.8, use its normal bundle given by $\{\nabla g_j\}_{j=1}^{\alpha}$; the gradients are evaluated at $F(\mathbf{R}^*)$ and α is the dimension of the the normal bundle.

---

[10]Simply stated, when four particles define Fig. 3.5c configurations, a dimension is lost.

With this change, if $\{g_j(\xi)\}_{j=1}^{\alpha}$ describes the constraints for configurations of a certain type, then $\mathbf{R}^*$ is a central configuration iff

$$\nabla \tilde{U}(F(\mathbf{R}^*)) = \lambda \nabla \tilde{I}(F(\mathbf{R}^*)) + \sum_{j}^{\alpha} \lambda_j g_j(F(\mathbf{R}^*)) \tag{3.12}$$

for scalars (Lagrange multipliers) λ_j. I prefer the alternative condition

$$\nabla \tilde{I}\tilde{U}^2 = \sum_{j=1}^{\alpha} \lambda_j \nabla g_j \tag{3.13}$$

that requires $\nabla \tilde{I}\tilde{U}^2$ to be in the appropriate normal bundle.

The α value, which is the number of independent constraints, is the difference between $\binom{N}{2}$ and the dimensions specified in Thm. 3.6; e.g., $\binom{N}{2} - (N-1) = \binom{N-1}{2}$ constraints of the $\xi_{i,j} + \xi_{j,k} - \xi_{i,k}$ type are needed for collinear configurations, $\binom{N}{2} - (2N-3) = \binom{N-2}{2}$ constraints involving degenerate tetrahedrons are required for the general coplanar configurations, and $\binom{N}{2} - (3N-6) = \binom{N-3}{2}$ constraints involving degenerate pentahedrons are necessary for the three-dimensional configurations. (The general case for a q-dimensional configuration requires $\binom{N-q}{2}$ constraints.)

The numbers of constraints for the remaining stratification components are easy to compute. By knowing what kinds of configurations correspond to each component, the nature of the constraints is easy to determined. For instance, the constraint defining a Fig. 3.5b configuration is a degenerate tetrahedron. A Fig. 3.5c configuration needs another constraint to capture the straight line: it is the $\xi_{1,4} + \xi_{4,2} - \xi_{1,2} = 0$ degenerate triangle condition.

3.3.1 The "configurational averaged length" ξ_{CAL}

The program suggested by Eq. 3.13 requires finding information about the various gradients. A direct computation shows that

$$\frac{\partial \tilde{I}\tilde{U}^2}{\partial \xi_{i,j}} = A m_j m_k \left[\frac{\tilde{U}}{2M\tilde{I}} \xi_{j,k} - \frac{1}{\xi_{j,k}^2} \right] \tag{3.14}$$

where $A = (2\tilde{I}\tilde{U})^{-1}$. Notice that the $\frac{\tilde{U}}{2M\tilde{I}}$ multiple agrees with the $\frac{U}{2I}$ multiple for central configurations (Eq. 2.16, page 40); the $2M$ value in $2M\tilde{I}$ cancels the $(2M)^{-1}$ term in the definition of $\tilde{I}$. Indeed,

$$\frac{\tilde{U}}{2M\tilde{I}} = \frac{\sum_{j<k} \frac{m_j m_k}{\xi_{j,k}}}{\sum_{j<k} m_j m_k \xi_{j,k}^2}. \tag{3.15}$$

Because $\frac{\tilde{U}}{2M\tilde{I}}$ is homogeneous of degree -3, it is convenient to treat this term as an "averaged inverse-cube" value of the $\xi_{j,k}$'s.[11] By doing so, Eq. 3.14 describes how the inverse-cube value of each $\xi_{j,k}$ differs from a weighted averaged inverse-cube length $[2M\tilde{I}/\tilde{U}]^{1/3}$: this $[2M\tilde{I}/\tilde{U}]^{1/3}$ term is used often enough to warrant assigning it a name.

Definition 3.2 *For* $\xi = (\xi_{12}, \ldots, \xi_{N-1,N}) \in \mathbb{R}^{\binom{N}{2}}$, *call*

$$\xi_{CAL} = [2M\tilde{I}/\tilde{U}]^{1/3} = [\frac{\sum_{j<k} m_j m_k \xi_{j,k}^2}{\sum_{j<k} \frac{m_j m_k}{\xi_{j,k}}}]^{1/3} \qquad (3.16)$$

the "configurational average length." (CAL)

The above comment about "averaged length" becomes more apparent by rewriting Eq. 3.14 as

$$\frac{\partial \tilde{I}\tilde{U}^2}{\partial \xi_{i,j}} = Am_j m_k [\frac{\xi_{j,k}}{\xi_{CAL}^3} - \frac{1}{\xi_{j,k}^2}] = Am_j m_k \xi_{j,k} [\frac{1}{\xi_{CAL}^3} - \frac{1}{\xi_{j,k}^3}]. \qquad (3.17)$$

This expression shows, for instance, that $\nabla \tilde{I}\tilde{U}^2 = \mathbf{0}$, which is an unconstrained critical point of $\tilde{I}\tilde{U}^2$, if and only if $\xi_{j,k} = \xi_{CAL}$ for all $j \neq k$. This structure makes it easy to show that, with the exception of equilateral central configurations, a central configuration must always have some leg lengths longer than ξ_{CAL}, while some others must be shorter.

Theorem 3.7 *The signs of the components of* $\nabla \tilde{I}\tilde{U}^2$ *at a central configuration are determined by whether a leg length is greater than, or less than,* ξ_{CAL}. *More precisely,*

$$\xi_{j,k} > \xi_{CAL} \text{ iff } \frac{\partial \tilde{I}\tilde{U}^2}{\partial \xi_{i,j}} > 0; \quad \xi_{j,k} < \xi_{CAL} \text{ iff } \frac{\partial \tilde{I}\tilde{U}^2}{\partial \xi_{i,j}} < 0. \qquad (3.18)$$

For equilateral central configurations, each leg length equals ξ_{CAL}. *For any other central configuration, some leg lengths are strictly smaller than* ξ_{CAL}, *and others are strictly larger.*

Proof: The proof of Eq. 3.18 is a direct computation that uses Eq. 3.17.

Because $\tilde{I}\tilde{U}^2$ is homogeneous of degree zero, the gradient $\nabla \tilde{I}\tilde{U}^2$ at point ξ is orthogonal to the line connecting ξ and the origin $\mathbf{0}$. As this line is in the positive orthant, all components of the direction vector are positive. Thus the orthogonality condition requires $\nabla \tilde{I}\tilde{U}^2$ to have positive and negative entries. With positive masses, the leg-length assertion follows. $\square$

---

[11]Recall from Chap. 2 that $\frac{U}{2I} = \frac{U}{R^2} = \frac{RU}{R^3}$ where RU is the configurational measure. Thus this term does capture an "inverse cubed" average of the spacing among particles.

3.3.2 Signs of gradients–coplanar configurations

In Eq. 3.17 and Thm. 3.7 we have described $\nabla \tilde{I} \tilde{U}^2$: the more difficult task is to compute and describe ∇g_j. For instance, if we adopt the currently popular approach by following the lead of Dziobek [18] where we first compute the volume of a tetrahedron, and then find the gradient, we are guaranteed to encounter complicated computations that will test the patience of the reader—and the author. Fortunately there is an easy way to determine certain qualitative properties of the ∇g gradients that hold independent of how the $g_j = 0$ constraints, which define a degenerate lower dimensional configuration, are selected. Consequently, while all of the results derived next can be obtained by using specific choices of constraints, no longer do we need to do so because the conclusions can be obtained in a much simpler fashion.

To explain the approach, select g_j so that $g_j > 0$ represents a higher dimensional configurations and $g_j < 0$ represents ξ variables that are not constructible objects: so, ∇g_j points in the direction of non-degenerate configurations; e.g., for $g = \xi_{1,2} + \xi_{2,3} - \xi_{1,3}$, $\nabla g = (1, 1, -1)$ is the directional change to convert the collinear configuration into a triangle.

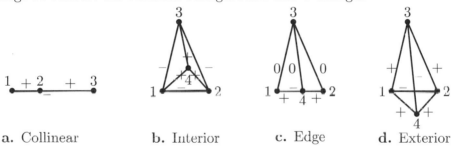

| a. Collinear | b. Interior | c. Edge | d. Exterior |

Fig. 3.6. Finding signs for gradients

Rather than carrying out the complicated task of computing the gradient, the idea is to determine the *signs* of the ∇g components by using the geometry and the definition of a partial derivative

$$\frac{\partial g}{\partial \xi_{j,k}} = \lim_{h \to 0} \frac{g(\xi_{1,2}, \ldots, \xi_{j,k} + h, \ldots, \xi_{N-1,N}) - g(\xi_{1,2}, \ldots, \xi_{j,k}, \ldots, \xi_{N-1,N})}{h}.$$

Using the partial—geometrically

By using the definition of the partial, the approach is to determine whether increasing, or decreasing, a leg length, while holding the others fixed, creates a two-dimensional configuration. For instance, in Fig. 3.6a, lengthening only the $\xi_{1,2}$ length rotates the remaining two legs about vertex 3 to create a triangle. This effect, where a larger $\xi_{1,2}$ creates a two-dimensional configuration, is denoted by the "+" in the figure: it reflects the sign of $\frac{\partial g}{\partial \xi_{1,2}} = 1$. Similarly, *shortening* $\xi_{1,3}$ causes a rotation about point 2 creating another triangle: this $\frac{\partial g}{\partial \xi_{1,3}} = -1$ sign is denoted by the "−" in the diagram.

An alternative approach is to determine what changes *cannot* be physically realized as leg lengths. Shrinking $\xi_{1,2}$, for instance, separates the segments at point 3: the new $\xi_{1,2}$ value violates the triangle inequality. Similarly, stretching $\xi_{1,3}$ forces the two short segments to separate at point 2. So, "squashing" lengths together corresponds to squeezing the configuration into a new dimension. "Tearing," on the other hand, destroys configurations and represents something that cannot be constructed. This alternative approach is needed to analyze degenerate four-dimensional objects.

To use this approach with the degenerate tetrahedron of Fig. 3.6b, notice that lengthening a leg connecting vertex 4 to one of the three vertices creates a tetrahedron; e.g., lengthening $\xi_{1,4}$ causes triangle $\{2, 3, 4\}$ to rotate along the $\{2,3\}$ edge: these legs are marked with "+'s." Alternatively, shortening $\xi_{1,4}$ is physically impossible because it would tear one of the rigid triangles $\{1, 2, 3\}$ or $\{4, 2, 3\}$, off of the $\{2, 3\}$ edge. The outside legs of the triangle are denoted by "−" because shortening these legs generates a tetrahedron.

Using this approach with Fig. 3.6d shows that we must assign negative signs to the two interior legs and positive signs to the four exterior legs. (For instance, a smaller $\xi_{1,2}$ would cause two triangles to rotate about the $\{3, 4\}$ edge, while a larger value would cause a tearing along this edge. Similarly, to avoid squashing the $\{1, 2, 3\}$ triangle into the $\{4, 2, 3\}$ triangle, a larger $\xi_{1,4}$ rotates two triangles along the $\{2, 3\}$ edge, while a smaller value would cause a tearing.) The signs for Fig. 3.6c are described later.

3.3.3   Signs of gradients–three-dimensional configurations

Three-dimensional configurations can be viewed as the degenerate form of a four-dimensional pentahedron as depicted in Fig. 3.7. To compensate for our weaker intuition about of the geometry of four-dimensional figures, the approach used for these configurations is to determine how a change in a leg length causes the three-dimensional figure to tear.

In this manner, we find that the three interior legs of Fig. 3.7a, given by dotted lines, must be assigned positive signs. This is because shortening one of these legs causes a tearing effect; e.g., shortening $\xi_{3,5}$ pulls the $\{5, 1, 2, 4\}$ tetrahedron off of the $\{1, 2, 4\}$ face. Similarly, the exterior legs given by the solid and dashed lines have negative gradient values; e.g., increasing

$\xi_{1,3}$ tears the $\{1,2,4,5\}$ tetrahedron from the $\{3,2,4,5\}$ tetrahedron along the common $\{2,4,5\}$ face. Incidentally, the choice of the tetrahedrons to examine is clear. For instance, when considering changes in the length of a $\xi_{1,4}$ leg, particle 1 must be on one of the tetrahedrons and particle 4 must be on the other: as there are only three remaining particles, the unique choices for this case are the $\{1,2,3,5\}$ and $\{4,2,3,5\}$ tetrahedrons.

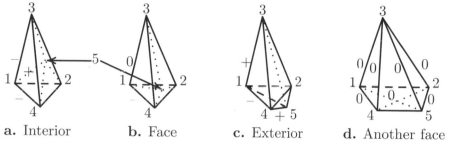

a. Interior   **b.** Face   **c.** Exterior   **d.** Another face

Fig. 3.7. Signs for three-dimensional gradients

Similarly, for Fig. 3.7c where vertex 5 has passed through the $\{1, 2, 4\}$ face (Fig. 3.7b), the only interior leg of $\xi_{3,5}$ (the dotted line) has a negative value: this is because a larger $\xi_{3,5}$ would cause a splitting on the $\{1,2,4\}$ face of tetrahedrons $\{1,2,4,5\}$ from $\{1,2,4,3\}$. The $\xi_{1,4}$, $\xi_{1,2}$, $\xi_{2,4}$ edges have negative values as a larger value would generate a tearing. (A larger $\xi_{1,2}$ value tears along the $\{3,4,5\}$ face forcing tetrahedrons $\{2,3,4,5\}$ and $\{1,3,4,5\}$ to separate.) The six remaining legs have positive signs. (For instance, a shorter $\xi_{1,3}$ separates on the $\{2,4,5\}$ face of tetrahedrons $\{1,2,4,5\}$ and $\{3,2,4,5\}$.) To summarize in a simpler manner, the sign of the sole interior diagonal and the three legs of the triangle *not* connected to the diagonal are negative, the signs of the six legs connecting endpoints of the diagonal to the triangle are positive.

3.3.4   Degenerate configurations

It remains to analyze the Fig. 3.6c configuration, where vertex 4 is on the line joining two other vertices, and the Figs. 3.7b, d configurations, where vertex 5 is on the plane passing through three other vertices. By comparing the signs for the "before" and "after" configurations—Figs. 3.6b and 3.6d—continuity mandates zero values for the three legs in Fig. 3.6c that connect vertex 3 with the three vertices on the line. This makes sense: stretching or shrinking $\xi_{j,3}$ while holding the others fixed create a non-constructible figure, so this component of ∇g should be zero. Similarly for Fig. 3.7b, the four legs connecting vertex 3 to the four vertices on the plane have zero values.[12] A caveat is necessary: rather than zero values, with some choices of a constraint g_j, these terms are limit values. With Fig. 3.6c, for instance, rather than $\frac{\partial g}{\partial \xi_{1,3}} = 0$, we may have $\frac{\partial g}{\partial \xi_{1,3}} / \frac{\partial g}{\partial \xi_{1,2}} \to 0$ as particle 4 approaches the bottom edge. (This happens with certain choices of g, but not with others.) Either case suffices for our purposes.

Notice how the sign pattern of the three vertices on the line in Fig. 3.6c agrees with the sign pattern of the collinear configuration, and the sign pattern of the six legs in the plane in Fig. 3.7b agrees with sign pattern of Fig. 3.6a. (The signs agree, the values need not.) This is no accident: it reflects the singularity structure of these configurations. For instance, the same argument shows that if vertex 5 in Fig. 3.7c keeps moving to be in the plane defined by vertices 2, 3, 4, then the sign structure of six legs would be that of Fig. 3.6d, while the legs connecting vertex 1 would have zero values. (Continuing to move vertex 5 makes $\xi_{1,5}$ the sole interior diagonal.)

The remaining and fascinating case is Fig. 3.7d. (Another case, which is left for the reader, has four particle in a plane defining a Fig. 3.6c configuration and the fifth out of the plane. Using the following, this setting is easy to analyze.) In Fig. 3.7d, particle 5 is at a transition point between having $\xi_{3,5}$ and $\xi_{2,5}$ serving as the sole interior diagonal: both will change signs with a change in particle 5, so at the Fig. 3.7d transition, the entry for both in ∇g is zero. Similarly, it could be particle 1, or 2, or 4 that defines a transition which could create one or another leg as the sole diagonal. By considering all of the combinations, it follows that the ∇g entry for all $\xi_{3,j}$ terms and the two diagonals in the base are zero.

It remains to consider the four legs that define the base. To do so, notice that if particle 5 moves downwards, to create $\xi_{3,5}$ as the sole interior diagonal, then $\{1, 2, 4\}$ is the triangle that is not attached to this diagonal: here these three legs have a negative value in the gradient, while the other six legs have positive values. If particle 5 moves upwards to create $\xi_{1,5}$ as the sole interior diagonal, then the triangle free from the diagonal is $\{2, 3, 4\}$, so now these legs have negative values in ∇g while the other six have positive values. The transition in sign means that $\xi_{1,4}, \xi_{1,2} = 0$ in the gradient. Using the same argument with other vertices in the base leads to the conclusion that $\nabla g = \mathbf{0}$—all legs are assigned the gradient value of zero!

---

[12]Readers interested in q-dimensional physical spaces for $q > 3$ can show that the same assertion holds. Namely, for a q dimensional configuration consisting of $q + 1$ particles where q particles are on a $(q - 1)$-dimensional hyperplane, the gradient of this constraint has zero entries for all legs that connect the particle that is off the hyperplane to the other particles.

3.4 Consequences for central configurations

The elementary arguments used to determine the sign patterns of the constraint equations provide several new insights into the structure of some central configurations. The approach is simple: up to a ± 1 common multiple, the ∇g and $\nabla \tilde{I} \tilde{U}^2$ components must have the same sign. These conclusions hold for any choice of a constraint g one wishes to use. Thus, all of the results described next can be obtained by using the reader's favored choice of a g constraint (where using a specified choice usually requires carrying out complicated computations), and some may have been. But this "rule of signs" approach holds for all choices and it is significantly simpler. A selection of immediate results follow.

3.4.1 Surprising regularity

All of the above configurations, Figs. 2.8b-d, 2.9a-d, seem to be reasonable candidates for central configurations—at least for some mass values. Even more, we might wonder whether the Fig. 3.6b four-body central configurations could be nearly collinear with vertices 1 and 2 close to one another, or with vertex 4 close to the line defined by 1 and 3. Similarly in Fig. 3.7a, could vertex 3 be near the plane defined by $\{1, 2, 4\}$, or could one of the $\{1, 2, 4\}$ legs be "small" relative to the length of a leg connecting vertex 5?

None of this can happen. As shown next the central configurations must be reasonably regular *for all choices of the masses.* For any (positive) mass values, a N-body central configurations that is $(N-2)$-dimensional is surprisingly regular.

Theorem 3.8 *The two short distances in a three-body collinear central configuration are strictly less than ξ_{CAL}, while the longer distance is strictly larger than ξ_{CAL}.*

For a four-body coplanar central configuration, it is impossible to have three particles along the same line. (So, configurations of the Fig. 3.6c type can never be central configurations.) If one particle in the interior of the triangle defined by the other three (that is, configurations of the Fig. 3.6b type), then the three leg lengths in the interior of the triangle are strictly less than the CAL value of ξ_{CAL} while the lengths of the exterior legs are strictly larger than ξ_{CAL}. If the four particles define a quadruple (a Fig. 3.6d type configuration), each of the four exterior leg lengths is strictly smaller than the CAL value, and each of the two diagonals is strictly larger.

If there exists a five-body non-coplanar central configurations where four of the particles are in a plane, then these four particles are on a circle and *the fifth particle is directly above the center of the circle. For other five-body non-coplanar central configurations, if one particle is in the interior of the hull defined by the other four (see Fig. 3.7a), the four interior leg lengths are strictly smaller than ξ_{CAL}; each of the remaining six leg lengths is strictly larger. If the five particles have no interior point (a Fig. 3.7c type configuration), then the lengths of the only interior diagonal and the three leg lengths that do not share a vertex with this diagonal are all greater than the CAL value; the remaining six leg lengths, on the exterior of the hull, are strictly smaller than the CAL value.*

Restating the theorem in kinder terms, each external leg of a central configuration of a Fig. 3.6b kind must be longer than each internal leg. Because the actual Fig. 3.6b has $\xi_{1,2} < \xi_{3,4}$, it cannot be a central configuration for any choice of the masses. Similarly, the theorem requires each diagonal of a Fig. 3.6d type central configuration to be longer than any of the exterior lengths. But as $\xi_{1,2} < \xi_{1,5}$ in Fig. 3.6d, it cannot be a central configuration for any choice of masses. These regularity conditions have several implications; e.g., they show, for instance, that a coplanar central configuration cannot be nearly a collinear configuration. Also, because Fig. 3.6c has three particles on a line, it cannot be a central configuration.

The same story extends to the Fig. 3.7 types of configurations: for Fig. 3.7a, each exterior leg length must be larger than each interior leg length (which is not true for this figure). To describe central configurations of the Fig. 3.7c type, place the sole diagonal on the z-axis. The lengths of the diagonal and the three legs that do not touch the z axis must be larger than any of the six legs that connect the triangle to the diagonal. Consequently, the figure cannot be too elongated; e.g., the actual Fig. 3.7c is not a central configurations because the three triangle legs are shorter than the other six. Similarly, it cannot overly squat with a short diagonal: thus, do not expect the three-dimensional five-body configurations to closely approximate two-dimensional ones.

As Fig. 3.7b has four particles in a plane, but one is in the interior of the triangle (so they are not on a circle), this configuration is not a central configuration for any choice of masses. As drawn, Fig. 3.7d is not a central configuration for any choice of masses because particle 3 is not directly above where the diagonals cross.

Proof: For a Fig. 3.6c or 3.7b, d configuration, all legs coming from one vertex have a zero value in ∇g. According to Eq. 3.13 (the Lagrange multiplier equation), if such a configuration is a central configuration, then each of the corresponding coordinates of $\nabla \tilde{I}\tilde{U}^2$ is zero. Thus, according to Thm. 3.7, each leg length equals ξ_{CAL}. We know from elementary geometry, that, for Figs. 3.6c, 3.7b, this is is impossible. This configuration, however, is possible for a Fig. 3.7d configuration if the particles in the plane are on a circle and the remaining particle is directly above the center. (This degenerate setting requires another constraint ensuring that particles 1, 2, 4, 5 form a degenerate tetrahedron. To satisfy this constraint, the particles must be on a circle and the diagonals must cross at the center. This condition has implications (not considered in this section) for the mass values.)[13]

A similar argument holds for the rest of the proof—the signs of ∇g, as given in the figures, determine whether the corresponding leg length of the central configuration is larger, or smaller, than ξ_{CAL}. This completes the proof. $\square$

Notice how the Lagrange multiplier λ in $\nabla \tilde{I}\tilde{U}^2 = \lambda \nabla g$ must always be negative in these settings. This is because, by choice, g was selected so that ∇g points toward a higher dimensional configuration. But $\tilde{I}\tilde{U}^2$ decreases in value when moving into higher dimensional settings. This difference causes a difference in signs. Consequently, when using the rule of signs with Figs. 3.6, 3.7 to compare leg lengths of a central configuration with ξ_{CAL}, remember that "negative" implies "longer."[14]

Negative masses

To suggest other applications of this rule of signs, suppose some masses values are negative. This causes no difficulties with the above analysis because the signs of the gradients of ∇g are independent of the masses: they reflect the physical constraints about how leg-lengths can, or cannot, be assembled to create a figure in different dimensional spaces. The signs of the masses, whether positive, zero, or negative, affect the signs of the components of $\nabla \tilde{I}\tilde{U}^2$—and the conclusions.

Negative masses! Why? Actually, negative masses are used to better understand the role of masses in celestial mechanics (e.g., see Gaerth Roberts [66] insightful paper on central configurations), or when electrical charges are included (e.g., a starting place is Perez, Saari, Suslin, and Yan [58], but most surely there are earlier and more recent papers on this topic). Using

---

[13]Similar assertions about $N - 1$ particles residing in a $N - 3$ dimensional hyperplane can be found for any dimension of "physical space" by using the fact described in the footnote on page 108,

[14]I considered selecting g to make the signs of ∇g and $\nabla \tilde{I}\tilde{U}^2$ agree at a central configuration. But, this choice causes confusion when using actual constraints; e.g., it would require a triangle to have a "negative area" and a tetrahedron to have a "negative volume."

the rule of signs in these settings leads to combinations where the $m_j m_k$ values are positive or negative and $\xi_{j,k}$ is larger than, or shorter than, ξ_{CAL}. Namely, if $x_{j,k}$ must be larger than ξ_{CAL} for a central configuration when $m_j m_k > 0$, it must be smaller than ξ_{CAL} when $m_j m_k < 0$. The following illustrates what can be found by using Fig. 3.6b: similar assertions follow for Figs. 3.6d and 3.7.

Corollary 3.1 *In Fig. 3.6, suppose that m_1 and m_2 have one sign, while m_3 has the other. If the sign of m_4 agrees with m_1, then a central configuration separates particle 3 from the other three in the sense that $\xi_{1,2}, \xi_{1,4}, \xi_{2,4} < \xi_{CAL}$ while all three leg lengths connecting vertex 3 are greater than ξ_{CAL}. But, if the sign of m_4 differs from that of m_1, then two binary systems, $\{1,2\}$ and $\{3,4\}$, are created in the sense that $\xi_{1,2}, \xi_{3,4} < \xi_{CAL}$, while the other four legs have length greater than ξ_{CAL}.*

Proof: The difference in using the rule of signs is that, for the first case, $m_1 m_2, m_1 m_4, m_2 m_4$ have positive values while the three $m_j m_3$ terms have negative values. Comparing the signs of $\nabla \tilde{I}\tilde{U}^2$ with those of Fig. 3.6b, the sign structures can be satisfied only if $\xi_{1,3}, \xi_{2,3}, \xi_{3,4}$ are on one side of ξ_{CAL} while the other three leg lengths are on the other. The choice is immediate because geometry requires that either $\xi_{1,3}$ or $\xi_{2,3}$ is longer than one of $\xi_{1,4}$ or $\xi_{2,4}$. The second assertion is proved in the same manner. $\square$

3.4.2   Estimates on ξ_{CAL}

So far we have not stated much about ξ_{CAL}. Even simple inequalities bounding this value provide other consequences about central configurations. For instance, Prop. 3.1 tells us that ξ_{CAL} reflects a mixture between the minimum and maximum spacing between particles given, respectively, by

$$\xi_{min} = \min_{j \neq k} \xi_{j,k}, \quad \xi_{max} = \max_{j,k} \xi_{j,k}.$$

As asserted next, ξ_{CAL} is roughly equal to $(\xi_{min} \xi_{max}^2)^{1/3}$.

Proposition 3.1 *For a given ξ and any choice of positive masses, there are two positive constants A and B depending on the masses so that*

$$A(\xi_{min} \xi_{max}^2)^{1/3} < \xi_{CAL} < B(\xi_{min} \xi_{max}^2)^{1/3} \tag{3.19}$$

Proof: If we let $m_0 = \min_j m_j$ and notice that at least one of the coordinates of ξ has the smallest value, say $\xi_{j,k} = \xi_{min}$, we obtain the inequality

$$\frac{m_0^2}{\xi_{min}} \leq \frac{m_j m_k}{\xi_{min}} \leq \tilde{U} \leq \frac{\sum_{j<k} m_j m_k}{\xi_{min}}.$$

In the same manner, it follows that

$$m_0^2 \xi_{max}^2 \leq \sum_{j<k} m_j m_k \xi_{j,k}^2 \leq \xi_{max}^2 \sum_{j<k} m_j m_k.$$

Equation 3.19 now follows from these estimates and the expression for ξ_{CAL} given in Def. 3.2 (page 104). $\sqcup$

While simple, Prop. 3.1 provides interesting information. For instance, notice that Thm. 3.8 does not preclude a Fig. 3.6b situation where $\xi_{1,4}$ and $\xi_{1,2}$ are nearly equal to ξ_{CAL} because the leg length $\xi_{2,4}$ becomes arbitrarily small: for this to occur, two of the particles must be near a collision! This cannot happen: the following theorem requires the minimum spacing between particles in a central configuration to be bounded away from zero.

Theorem 3.9 *For any central configuration of the Figs. 3.6b,d, 3.7a,c types, there is a positive bound A, which depends on the masses and type of configuration, so that*

$$A\xi_{max} \leq \xi_{min}. \tag{3.20}$$

Proof: The argument is given for a Fig. 3.6b type configuration: only minor modifications are needed for the other configurations. First, notice that a combination between the triangle inequality and Thm. 3.8 require the inequality $\frac{1}{2}\xi_{max} < \xi_{CAL}$. (This is because the interior leg lengths are smaller than ξ_{CAL}, but at least two of these legs form a triangle with the exterior leg that has length ξ_{max}.) This inequality with Eq. 3.19 leads to

$$\frac{1}{2}\xi_{max} < \xi_{CAL} < B\xi_{min}^{1/3}\xi_{max}^{2/3}.$$

The extreme ends of this inequality show that

$$\frac{1}{2B}\xi_{max}^{1/3} < \xi_{min}^{1/3}.$$

The conclusion follows. $\square$

3.4.3   Are there central configurations of these types?

The rule of signs, with results of the type as given by Thm. 3.8, describes conditions on the shape of central configurations in terms of constraints on the leg-lengths. But, Thm. 3.8 does not answer a fundamental concern: are there any central configurations of these Figs. 3.6, 3.7 types? The rule of signs allows us to prove that there are.

Theorem 3.10 *For any choices of positive masses, there always exist copla-nar four-body central configurations of the following types: any three of the particles define a non-degenerate triangle and the fourth is in the interior; or, no particle is in the hull defined by the other three.*

For any choices of positive masses, there always exist non-coplanar five-body central configurations of the following types: any four of the particles define a tetrahedron, and the fifth is in its interior; or, no particle is in the interior of the non-degenerate tetrahedron defined by the other four.

Proof: The proof follows immediately from standard calculus. Again, the proof is provided for Fig. 3.6b kinds of configurations as the details are essentially the same for all configurations.

With $\tilde{I} = 1$, let $\mathcal{C}$ be the set of ξ that define a degenerate tetrahedron where particle 4 is in the interior of the hull defined by the other three particles so that

1. $\xi_{j,4} \leq \xi_{CAL}$, but $\xi_{j,k} \geq \xi_{CAL}$, $j, k = 1, 2, 3$, while

2. $\frac{A}{2}\xi_{max} \leq \xi_{min}$ where A is defined in Eq. 3.20.

As any central configuration of this type must satisfy these two condi-tions, these constraints do not eliminate any possibilities. These conditions are imposed to create a compact set over which $\tilde{I}\tilde{U}^2$ is defined and smooth. For instance, condition 2 keeps us away from settings where some $\xi_{i,j} = 0$: this is where $\tilde{U}$ is not defined. This restriction causes no difficulties because, as we now know from Thm. 3.9, no central configuration can occur in this region. Thus conditions 1 and 2 ensure that $\mathcal{C}$ is a compact set. In turn, this means that $\tilde{I}\tilde{U}^2$ has a critical point over $\mathcal{C}$.

Physical considerations prohibit all legs from being of length ξ_{CAL} (i.e., this figure cannot be an equilateral configuration), so some components of $\nabla \tilde{I}\tilde{U}^2$ must be non-zero. Because of condition 1, the sign of each of the components of $\nabla \tilde{I}\tilde{U}^2$ is the opposite of the sign of the same component of ∇g. Thus, it follows that $\tilde{I}\tilde{U}^2$ has a smaller value in the interior of $\mathcal{C}$. This interior critical point is the central configuration. $\square$

3.5   What can, and cannot, be

Using the rule of signs with more particles requires using more constraints. Five particles, for instance, define ten leg lengths. A three dimensional five-body configuration has nine degrees of freedom, so its SMD(10) subspace can be described with a single degenerate pentahedron. But for five-body coplanar configurations, the number of degrees of freedom is at most seven (Thm. 3.6), so three "degenerate tetrahedron constraints" are needed.

It might appear from Eq. 3.13 that the more complex settings require using actual ∇g_j entries—unless certain $\xi_{j,k}$ entries are zero for most ∇g_j constraints. This observation suggests examining configurations constructed with Fig. 3.6c or 3.7b, d components. By doing so, the rule of signs leads to conclusions that, again, demonstrate the regularity of the central configurations. To illustrate the kinds of results that will occur, I use Fig. 3.8

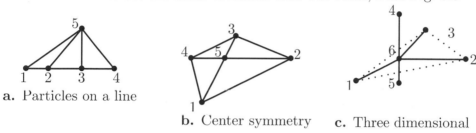

a. Particles on a line

b. Center symmetry   **c.** Three dimensional

Fig. 3.8. Composite configurations

3.5.1   More central configurations

While lessons learned from Fig. 3.6c suggest that Fig. 3.8a cannot be a central configuration, it is reasonable to suspect that Figs. 3.8b, c are central configurations for carefully selected mass choices. (Figure 3.8b is a coplanar configuration, Fig. 3.8c is a three-dimensional configuration.) But, as proved below, as drawn, none is a central configuration for any positive mass choices. For instance, if Fig. 3.8a were a central configuration, then the lengths of all legs connecting a particle with particle 5 must equal ξ_{CAL}: this is impossible. The explanations for the other configurations, described below, exploit certain symmetries of the configurations.

Five-body configurations of the Fig. 3.8a type are characterized by a five-dimensional SMD($\binom{5}{2}$) subspace; e.g., knowing the configuration, it suffices to know $\xi_{1,2}, \xi_{2,3}, \xi_{3,4}, \xi_{1,5}$ and $\xi_{4,5}$. Thus $\binom{5}{2} - 5 = 5$ constraints are needed: three independent constraints describe degenerate tetrahedrons, and two describe the degenerate triangles that force three particles on a line segment. The only constraints involving particle 5 come from the three degenerate tetrahedron conditions: the signs of their gradient structures are as follows (where a blank means the variable is not involved, so it corresponds to a 0):

	$\xi_{1,2}$	$\xi_{1,3}$	$\xi_{1,4}$	$\boldsymbol{\xi_{1,5}}$	$\xi_{2,3}$	$\xi_{2,4}$	$\boldsymbol{\xi_{2,5}}$	$\xi_{3,4}$	$\boldsymbol{\xi_{3,5}}$	$\boldsymbol{\xi_{4,5}}$
$\{1,2,3,5\}$	+	−		0	+		0		0	
$\{1,2,4,5\}$	+		−	0		+	0			0
$\{2,3,4,5\}$					+	−	0	+	0	0
Sum				0			0		0	0

$$(3.21)$$

To establish the linear independence of the Eq. 3.21 vectors, notice that the $\xi_{1,3}, \xi_{1,4}, \xi_{3,4}$ components are non-zero for precisely one gradient. "Independence" follows from the placement of these components.

Key for our argument is that, independent of the choices of the Lagrange multipliers in Eq. 3.13, the sum of each $\xi_{j,5}$ component in Eq. 3.21 is zero. Therefore, if such a configuration were a central configuration, then $\frac{\partial \tilde{I}\tilde{U}^2}{\partial \xi_{j,5}} = 0$ for $j = 1, 2, 3, 4$. According to Thm. 3.7, each of these leg lengths equals ξ_{CAL}, which is impossible.

Theorem 3.11 *A coplanar central configuration for $N \geq 4$ never has $N-1$ of the particles on a line and one off of it. Similarly, if a non-coplanar central configuration for $N \geq 5$ exists with $N-1$ of the particles in a plane and one off of it, then the $N-1$ particles are on a circle where the final particle is directly above the center of the circle.*

Consider five-body coplanar configurations of the Fig. 3.8b type. Namely, start with a quadruple defined by four particles, and place the fifth where the two diagonals cross. For such a configuration to be a central configuration, the lengths of the external legs connecting adjacent particles (the edges of the quadruple) are all greater, equal, or smaller than ξ_{CAL}.

For six-body configurations of the Fig. 3.8c type, start with a degenerate pentahedron that has a single diagonal. Place the sixth particle at the point where this diagonal crosses the plane defined by the three particles that are not on the diagonal. If such a configuration is a central configuration, the lengths of the six external legs that that meet one of the ends of the diagonal are all greater, equal, or smaller than ξ_{CAL}.

The assertion about a $N-1$ particles on a line most surely is somewhere in the literature, but I did not find it: the other assertions appear to be new.[15] While more ambitious results can be found by using this rule of signs technique, I have not pursued this direction in any serious, systematic

---

[15]We could, of course, analyze what happens with a mixture between positive and

manner. Care is needed; e.g., the geometry and the fact that at least one $\nabla \tilde{I} \tilde{U}^2$ component is negative shorter than ξ_{CAL}. After all, ξ_{CAL} depends on the configuration and masses: as the fifth body permits shorter $\xi_{j,5}$ lengths, ξ_{CAL} might be smaller than the exterior leg lengths.

Proof: Configurations with $N-1$ of the N particles on a line define a N-dimensional $\mathrm{SMD}(\binom{N}{2})$ submanifold: let m_N be the particle off the line. The defining constraints for this space are a combination of conditions describing degenerate tetrahedrons and degenerate triangles (for triplets on the line). All constraints involving $\xi_{j,N}$ for any $j = 1, \ldots, N-1$ describe tetrahedrons that, by the geometric assumption, are of the Fig. 3.6c type. Thus, for each j, the $\xi_{j,N}$ component of ∇g for these constraints is zero. If such a configuration were a central configuration, it follows from the Lagrange multiplier formulation in Eq. 3.13 and Thm. 3.7 that all legs would equal ξ_{CAL}. This is geometrically impossible, so the conclusion follows.

The proof for a three-dimensional configuration, where $N-1$ of the particles are in a plane and m_N is out of the plane, is essentially the same. All constraints involving $\xi_{j,N}$ variables are degenerate pentahedrons of the Fig. 3.7b or d type. The same argument using Thm. 3.7 shows that if such a configuration were a central configuration, each $\xi_{j,N}$ length would equal ξ_{CAL}. Such a configuration does not exist if it is of the Fig. 3.7b type. But, it could exist if all $N-1$ particles were on a circle and the Nth particle is directly above the center. (Restrictions on the particles in the plane are determined by degenerate tetrahedron constraints.) The conclusion follows.

Now consider a five-body configuration of the Fig. 3.8b type where m_5 is the interior particle. These configurations define a six-dimensional $\mathrm{SMD}(\binom{5}{2})$ submanifold: the defining constraints are three independent conditions involving degenerate tetrahedrons and "straight line" constraints that force m_5 on each of the two diagonals. The external leg variables occur only in the degenerate tetrahedron conditions. Four of these tetrahedron conditions involve m_5: each of them is of the Fig. 3.6b type where the value of the gradient components for an exterior leg is zero. But, it is not clear from the signs of entries whether the gradients of these four constraints span a three-dimensional space (they do not), so the gradient for the remaining degenerate tetrahedron $\{1, 2, 3, 4\}$ must be used. This constraint is of a Fig. 3.6d type where the entries for all external legs have the same sign. Because this is the only constraint involving external legs, their components in $\nabla \tilde{I} \tilde{U}^2$

---

negative mass values. In this manner we find, for instance, that the assertion about $N-1$ particles being on a line, or a plane, still holds; or a central configuration of a Fig. 3.8d form with orthogonal diagonals can occur only if masses on end points of a diagonal have the same sign.

must all have the same sign, or all be zero if the Lagrange multiplier for this constraint is zero. The conclusion now follows from Thm. 3.7.

The six-body Fig. 3.8c configurations define a seven-dimensional $\mathrm{SMD}(\binom{6}{2})$ submanifold defined by eight degenerate pentahedrons constraints, one degenerate tetradron (involving vertices 1, 2, 3, 6) and a degenerate triangle (involving vertices 4, 5, 6). A new feature is that the sign structures of some of the $\binom{6}{5} = 6$ degenerate pentahedrons are not given by Fig. 3.7. For instance, the pentahedron $\{1, 2, 4, 5, 6\}$ in Fig. 2.10, where three vertices are on an edge, is not represented in Fig. 2.9. This sign structure given in Fig. 3.9, however, follows immediately from continuity and Figs. 3.7a, c.

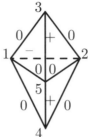

Fig. 3.9. Signs for another three-dimensional gradient

To explain, in Fig. 3.7a, move vertex 5 toward the $\xi_{3,4}$ edge. As long as vertex 5 is in the interior of the tetrahedron, the signs associated with $\xi_{j,5}$ partials are positive, and all other partials are negative. After vertex 5 passes through the $\xi_{3,4}$ edge to create a object like Fig. 3.7c with the only diagonal is $\xi_{3,4}$, the signs of all but four partials change. The ones that remain the same are for $\xi_{3,5}, \xi_{4,5}$ (they are positive) and $\xi_{1,2}, \xi_{3,4}$ (they are negative). Continuity determines the sign structure for the degenerate Fig. 3.9 as indicated. Namely, the edge with three vertices has the sign structure of a degenerate triangle. Of the remaining seven variables, six have a zero partial derivative; the non-zero (negative) partial derivative is for the $\xi_{1,2}$ variable representing the leg with no vertices on the edge.

Six pentahedrons that can be constructed from Fig. 3.8c: $\{1, 2, 3, 4, 6\}$ and $\{1, 2, 3, 5, 6\}$) are of the Fig. 3.7b type, $\{1, 2, 4, 5, 6\}$, $\{1, 3, 4, 5, 6\}$, and $\{2, 3, 4, 5, 6\}$ are of the Fig. 3.9 type with three particles on the center diagonal, and $\{1, 2, 3, 4, 5\}$ is of the Fig. 3.7c type. All other constraints involve the center straight line and the $\{1, 2, 3, 6\}$ tetrahedron. For the first five gradients, then, the $\xi_{j,4}$ and $\xi_{j,5}$, $j = 1, 2, 3$, partials are zero. In the last gradient of a pentahedron, they all have the same sign. Thus, for the configuration to be a central configuration, all of these leg lengths are greater, or less, than ξ_{CAL} if the Lagrange multiplier is non-zero. Otherwise, all of these leg lengths equal ξ_{CAL}. This completes the proof. $\square$

3.5.2 Masses and collinear central configurations

A lesson learned from the rule of signs is that central configurations reflect interesting differences between the structures of ∇g and $\nabla \tilde{I}\tilde{U}^2$. The ∇g's depend only on the geometry of the SMD subspaces: a geometric singularity caused by too many particles on a straight line or plane is manifested by zeros in certain gradient components, or, as in the collinear three-body problem, with ± 1 entries independent of the distances. As we know from introductory calculus, Lagrange multiplier problems must involve the constraints: consequently, to find central configuration results, at some time we must appeal to the basic geometry. The $\nabla \tilde{I}\tilde{U}^2$ entries, on the other hand, ignore the geometry: they are sensitive only to mutual distances *and* mass values. So, it is $\nabla \tilde{I}\tilde{U}^2$ that determines how central configurations change with changes in mass values.

To illustrate these differences with a Fig. 3.6b configuration (page 105) with a particle inside a triangle, if the center particle approaches a vertex its component in the ∇g constraint may decrease in magnitude, but the $\nabla \tilde{I}\tilde{U}^2$ component must increase. Thus, such a configuration can be a central configuration only if that vertex has a sufficiently small mass value.[16] In this subsection I examine how mass values affect central configurations.

A natural starting place is with collinear three-body configurations because the $\nabla(\xi_{1,2}+\xi_{2,3}-\xi_{1,3}) = (1, 1, -1)$ gradient for the SMD(3) constraint ignores distances. (As in Fig. 3.6a, particle 2 is in the middle.) Combining this gradient structure with $\nabla \tilde{I}\tilde{U}^2$, where the Eq. 3.17 components

$$\frac{\partial \tilde{I}\tilde{U}^2}{\partial \xi_{j,k}} - Am_j m_k \left[\frac{\xi_{i,j}}{\xi_{CAL}^3} - \frac{1}{\xi_{j,k}^2} \right] \tag{3.22}$$

are reproduced here for convenience, a collinear configuration must balance position and mass values in locating particle 2 to create the equality

$$\frac{\partial \tilde{I}\tilde{U}^2}{\partial \xi_{1,2}} = \frac{\partial \tilde{I}\tilde{U}^2}{\partial \xi_{2,3}} = -\frac{\partial \tilde{I}\tilde{U}^2}{\partial \xi_{1,3}} \tag{3.23}$$

If $m_1 = m_3$, so $m_1 m_2 = m_2 m_3$, it follows from simple algebra, Eq. 3.22, and the first two terms of Eq. 3.23 that $\xi_{1,2} = \xi_{2,3}$: the second particle must be at the midpoint between m_1 and m_3. Similarly, if $m_1 > m_3$, then Eq. 3.23 requires that $\xi_{2,3} < \xi_{1,2}$: particle 2 is closer to 3 than to 1.

---

[16]As it actually happens, certain ∇g components may grow, or decrease, so rapidly that no mass choices allow a configuration to be a central configuration.

*For three-body collinear central configurations, the middle parti-
cle is closer to the end particle with the smaller mass. If both end
masses are equal, the middle particle is located at the midpoint.*

Conversely, fixing the position of the end particles for the collinear three-
body problem defines an open set of possible locations for the middle particle: the problem is to find the unique position. It is interesting, and easy
to prove, that for each position of the middle particle, mass values can be
found to make this position a central configuration. This result follows immediately from continuity and the fact that as either $m_1 \to 0$ or $m_3 \to 0$
while the other mass has a fixed value, the associated central configuration
requires the middle particle to be arbitrarily close to the small mass.

Fig. 3.10. Collinear four-body problem

Analyzing $N > 3$ particles on the line uses a similar approach; the
notions are described for $N = 4$. The $(N-1)$-dimensional linear subspace
of $\mathrm{SMD}(\binom{N}{2})$ requires $\binom{N-1}{2}$ independent degenerate triangle constraints,
so, for $N = 4$, use any three of the four possible choices. Using the Fig. 3.10
ordering, the four gradient structures are:

Triplet	Weight	$\xi_{1,2}$	$\xi_{1,3}$	$\xi_{1,4}$	$\xi_{2,3}$	$\xi_{2,4}$	$\xi_{3,4}$
$\{1,2,3\}$	λ_1	1	-1		1		
$\{2,3,4\}$	λ_2				1	-1	1
$\{1,3,4\}$	λ_3		1	-1			1
$\{1,2,4\}$	λ_4	1		-1		1	

(3.24)

(The last row is the sum of the first and third minus the second.)

As Eq. 3.24 demonstrates, distances between particles play no role in the
constraints: they are partially captured by the Lagrange multipliers. Thus,
as with the collinear three-body problem, the $\nabla \tilde{I}\tilde{U}^2$ structure determines the
configuration where smaller mass values compensate for closer distances that
cause a larger $\xi_{j,k}^{-3}$ value. The proof of the next statement, which describes
certain natural settings, indicates how to use this observation. The idea for
the proof is to choose the constraints so that appropriate $\xi_{j,k}$ terms occur
in only one of the selected degenerate triangle constraints.

Theorem 3.12 *a. For a four-body collinear central configuration, with particles ordered as 1, 2, 3, 4 and for any positive masses, distances between adjacent particles are smaller than ξ_{CAL}; i.e., $\xi_{1,2}$, $\xi_{2,3}$, $\xi_{3,4} < \xi_{CAL}$.*

b. If the configuration has $\frac{\partial \tilde{I}\tilde{U}^2}{\partial \xi_{1,2}} = \frac{\partial \tilde{I}\tilde{U}^2}{\partial \xi_{3,4}}$, then $\frac{\partial \tilde{I}\tilde{U}^2}{\partial \xi_{1,3}} = \frac{\partial \tilde{I}\tilde{U}^2}{\partial \xi_{24}}$. If in addition $m_1 m_2 = m_3 m_4$, then $m_1 = m_4$, $m_2 = m_3$, and the central configuration has $\xi_{1,2} = \xi_{3,4}$. The length of the remaining $\xi_{2,3}$ depends on the ratio $m = m_2/m_1$. More precisely, setting $\alpha = \xi_{2,3}/\xi_{1,2}$,

$$m = \frac{\frac{1}{\alpha} - \frac{1}{\alpha(1+\alpha)^2} + \frac{2}{(2+\alpha)^3}}{\frac{1}{\alpha^3} - \frac{1}{2+\alpha} - \frac{1}{(2+\alpha)(1+\alpha)^2}} = \frac{\alpha^3[(2+\alpha)^4 + (1+\alpha)^2]}{(2+\alpha)^2[2 + 5\alpha + 4\alpha^2 - \alpha^3(1+\alpha)^2]}$$

$$(3.25)$$

The first part of Thm. 3.12 captures the regularity of four-body collinear configurations: by using the same approach (i.e., a judicious choice of the constraint equations as described in the proof), related results follow for any N. While the second part of this theorem stresses a special case, all settings can be similarly analyzed.

The special case provides insight into the central configuration structure. According to Eq. 3.25, a central configuration could have the two center particles arbitrarily close to each other—but only if their common mass value is sufficiently small. This assertion is consistent with the need to use smaller masses to counter larger $\xi_{2,3}^{-3}$ values. On the other hand, even with an arbitrarily large mass value for the two center particles, there is a limit to how distant the outside particles can be separated relative to the distance between the center ones. As described next, the outside legs ($\xi_{1,2}$ and $\xi_{3,4}$) can never as large as $3/2$ times the $\xi_{2,3}$ distance between the center particles. The following statement, which describes these and other the mass-distance interactions, is an immediate consequence of Eq. 3.25.

Corollary 3.2 *For all collinear four-body central configurations ordered as in Fig. 3.10 and with $m_1 = m_4$, $m_2 = m_3$, the two end distances are equal; i.e., $\xi_{1,2} = \xi_{3,4}$. For the remaining distance between adjacent particles, $\xi_{2,3}$, the ratio $\xi_{2,3}/\xi_{1,2}$ is restricted to*

$$0 < \xi_{2,3}/\xi_{1,2} < \alpha^* < 1.45 \qquad (3.26)$$

where α^* is the sole positive root of $2 + 5\alpha + 4\alpha^2 - \alpha^3 - 2\alpha^4 - \alpha^5 = 0$. The associated masses for a central configuration must have $m_2/m_1 \to 0$ iff $\xi_{2,3}/\xi_{1,2} \to 0$ and $m_2/m_1 \to \infty$ iff $\xi_{2,3}/\xi_{1,2} \to \alpha^*$. Indeed, the $\xi_{2,3}/\xi_{1,2}$ value increases monotonically from 0 to α^* as m_2/m_1 increases monotonically from 0 to ∞.

For a four-body collinear central configuration with equal distances between adjacent particles (so $\xi_{1,2} = \xi_{2,3} = \xi_{3,4}$) with $m_1 = m_4$, $m_3 = m_4$, the masses must satisfy $m_3/m_1 = \frac{85}{63} \approx 1.349206$. Thus, in the equal mass case of $m_1 = m_2 = m_3 = m_4$, the collinear central configuration has $\xi_{1,2} = \xi_{3,4} > \xi_{2,3}$.

All results, except the monotonicity assertion, are direct consequences of Eq. 3.25. The monotonicity assertion follows from a logarithmic differentiation of the right hand side of Eq. 3.25 to show that $\frac{dm}{d\alpha} > 0$ in $(0, \alpha^*)$.

Proof of Thm. 3.12: To prove that $\xi_{1,2} < \xi_{CAL}$, assume the contrary and use the first three gradients of Eq. 3.24 (as this creates a setting where only one $\xi_{1,2}$ component of a gradient is non-zero). The assumption that $\xi_{1,2} \geq \xi_{CAL}$ holds if and only if $\frac{\partial \tilde{I}\tilde{U}^2}{\partial \xi_{1,2}} \geq 0$, or iff $\lambda_1 \geq 0$. As the geometry requires $\xi_{1,3} > \xi_{1,2} \geq \xi_{CAL}$, it must be (Eq. 3.17) that $\frac{\partial \tilde{I}\tilde{U}^2}{\partial \xi_{1,3}} > 0$, or, from Eq. 3.24, that $-\lambda_1 + \lambda_3 > 0$; i.e., $\lambda_3 > 0$. But the $\xi_{1,4}$ component is non-zero in only gradient constraint (the third), so $\lambda_3 > 0$ requires $\frac{\partial \tilde{I}\tilde{U}^2}{\partial \xi_{1,4}} < 0$; this proves the assertion as it implies the impossible conclusion that $\xi_{1,4} < \xi_{CAL} < \xi_{1,3}$. It follows immediately from the symmetry between $\xi_{1,2}$ and $\xi_{3,4}$—both are the "outside" intervals— that because $\xi_{1,2}$ must be smaller than ξ_{CAL}, then $\xi_{3,4} < \xi_{CAL}$. Indeed, the proof is the same by reversing the names of the particles (i.e., a rotation of the configuration), or by using the first, second, and fourth rows.

To prove that $\xi_{2,3} < \xi_{CAL}$, use the first, third, and four rows of Eq. 3.24, and assume that $\xi_{2,3} \geq \xi_{CAL}$. The assumption requires $\lambda_1 \geq 0$. But because $\xi_{1,3} > \xi_{2,3}$, it must be that $-\lambda_1 + \lambda_3 > 0$, or $\lambda_3 > 0$. In turn, this means that the $\xi_{3,4}$ component of $\nabla \tilde{I}\tilde{U}^2$ equals λ_3, or that $\xi_{3,4} > \xi_{CAL}$: this contradiction completes the proof.

To prove part b, use the first three rows of Eq. 3.24. If $\frac{\partial \tilde{I}\tilde{U}^2}{\partial \xi_{1,2}} = \frac{\partial \tilde{I}\tilde{U}^2}{\partial \xi_{3,4}}$, then the gradient structure requires

$$\lambda_1 = \lambda_2 + \lambda_3. \tag{3.27}$$

Equation 3.24 further requires $\frac{\partial \tilde{I}\tilde{U}^2}{\partial \xi_{1,3}} = -\lambda_1 + \lambda_3$ and $\frac{\partial \tilde{I}\tilde{U}^2}{\partial \xi_{2,4}} = -\lambda_2$. Because Eq. 3.27 requires $-\lambda_1 + \lambda_3 = -\lambda_2$, we have that $\frac{\partial \tilde{I}\tilde{U}^2}{\partial \xi_{1,3}} = \frac{\partial \tilde{I}\tilde{U}^2}{\partial \xi_{24}}$.

By assumption $m_1 m_2 \xi_{1,2}[\frac{1}{\xi_{CAL}^3} - \frac{1}{\xi_{1,2}^3}] = m_3 m_4 \xi_{3,4}[\frac{1}{\xi_{CAL}^3} - \frac{1}{\xi_{3,4}^3}]$, so if $m_1 m_2 = m_3 m_4$, then it follows from algebra that $\xi_{1,2} = \xi_{3,4}$; i.e., the two end segments have equal length. These equal lengths on a collinear configuration require $\xi_{1,3} = \xi_{1,2} + \xi_{2,3} = \xi_{3,4} + \xi_{2,3} = \xi_{2,4}$. Using this equality with the just proved $\frac{\partial \tilde{I}\tilde{U}^2}{\partial \xi_{1,3}} = \frac{\partial \tilde{I}\tilde{U}^2}{\partial \xi_{24}}$, we have that $m_1 m_3 = m_2 m_4$. When combined with the $m_1 m_2 = m_3 m_4$ assumption, it follows from algebra that $m_1 = m_4$, $m_2 = m_3$.

There are several ways to establish Eq. 3.25. First notice that if all distances are described in terms of multiples of $\xi_{1,2}$ and all masses as multiples of m_1, then Eq. 3.22 has a common $A m_1^2 / \xi_{1,2}^3$ multiple and all other entries are m or α. By being common to all entries, this $A m_1^2 / \xi_{1,2}^3$ multiple can be absorbed in the Lagrange multipliers. In this manner, by using the first three rows of Eq. 3.24 and selecting an entry in each row that is the only non-zero term in that column,[17] we have that

$$\lambda_1 = m\left(\frac{1}{\xi_{CAL}^3} - 1\right), \ \lambda_2 = m\left(\frac{1}{(1+\alpha)^2} - \frac{1+\alpha}{\xi_{CAL}^3}\right), \ \lambda_3 = \frac{1}{(2+\alpha)^2} - \frac{2+\alpha}{\xi_{CAL}^3}.$$

The $\xi_{2,3}$ column of Eq. 3.24 requires

$$\lambda_1 + \lambda_2 = m^2\left(\frac{\alpha}{\xi_{CAL}^3} - \frac{1}{\alpha^2}\right),$$

which, with the λ_1, λ_2 values, leads to the central configuration expression

$$\frac{1}{\xi_{CAL}^3} = \frac{1}{(1+m)\alpha}\left[\frac{m}{\alpha^2} - 1 + \frac{1}{(1+\alpha)^2}\right]$$

Similarly, the $\xi_{1,3}$ column of Eq. 3.24 requires that

$$-\lambda_1 + \lambda_3 = m\left[\frac{1+\alpha}{\xi_{CAL}^3} - \frac{1}{(1+\alpha)^2}\right],$$

which leads to another central configuration expression

$$\frac{1}{\xi_{CAL}^3} - \frac{1}{(1+m)(2+\alpha)}\left[m + \frac{m}{(1+\alpha)^2} + \frac{1}{(2+\alpha)^2}\right].$$

Setting the ξ_{CAL}^{-3} expressions equal and solving for m leads to Eq. 3.25. $\square$

---

[17]This strategy, which is used throughout this proof in different ways, can be used with any $N \geq 4$. For $N = 5$, out of the $\binom{5}{3} = 10$ degenerate triangles, only six need be used: thus the strategy is to make choices where one Lagrange multiplier describes particular $\frac{\partial \tilde{I}\tilde{U}^2}{\partial \xi_{j,k}}$ values. For $N = 5$, I suggest listing all ten gradients as in Eq. 3.24. For $N > 5$, however, the larger number of mutual distances makes it more efficient to use an indirect argument rather than listing the gradients.

3.5.3   Masses and coplanar configurations

Similar arguments hold for coplanar and non-coplanar configurations. To suggest the kinds of arguments that can be used, a special case is described. In doing so, recall the five-body setting (page 116) where four of the particles are on a circle in a plane where the diagonals must meet at the center of the circle. If four particles were to form a central configuration of this type, are there any restrictions on the masses? As shown next, there are, and they follow from the natural symmetries that are captured by Fig. 3.11.

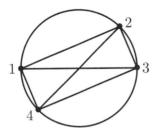

Fig. 3.11. Symmetries

The setting depicted in Fig. 3.11 has the two interior diagonals intersecting at the center of the circle where all four particles are on the circle. Elementary geometry requires the following:

1. The two diagonals of the configuration are diagonals of the circle, so they have equal length; e.g., $\xi_{1,3} = \xi_{2,4}$.

2. Because the diagonals pass through the center of the circle, the particles form a rectangle. Consequently, $\xi_{1,4} = \xi_{2,3}$ and $\xi_{3,4} = \xi_{1,2}$.

As the mass values play no role in the ∇g entries, the symmetry of this setting dictates that

$$\frac{\partial g}{\partial \xi_{2,4}} = \frac{\partial g}{\partial \xi_{1,3}}, \quad \frac{\partial g}{\partial \xi_{1,4}} = \frac{\partial g}{\partial \xi_{2,3}}, \quad \frac{\partial g}{\partial \xi_{1,2}} = \frac{\partial g}{\partial \xi_{3,4}} \qquad (3.28)$$

From these relationships, the following result is immediate.

Theorem 3.13 *If a four-body central configuration has all particles on a circle where the two connecting diagonals intersect at the center of the circle, then all masses are equal and the configuration is a square.*

Clearly, a square is one choice for a central configuration. That it is the only choice does not seem to follow directly from the simple symmetries and qualitative argument used here, but it does follow from a symmetry argument that uses the definition of a central configuration.

Proof: According to Eqs. 3.28, 3.17, and the Lagrange multiplier relationship, the three relationships in Eq. 3.28 imply that

$$
\begin{aligned}
Am_2m_4\left[\frac{\xi_{2,4}}{\xi_{CAL}^3} - \frac{1}{\xi_{2,4}^2}\right] &= Am_1m_3\left[\frac{\xi_{1,3}}{\xi_{CAL}^3} - \frac{1}{\xi_{1,3}^2}\right], \\
Am_1m_4\left[\frac{\xi_{1,4}}{\xi_{CAL}^3} - \frac{1}{\xi_{1,4}^2}\right] &= Am_2m_3\left[\frac{\xi_{2,3}}{\xi_{CAL}^3} - \frac{1}{\xi_{2,3}^2}\right], \qquad (3.29) \\
Am_1m_2\left[\frac{\xi_{1,2}}{\xi_{CAL}^3} - \frac{1}{\xi_{1,2}^2}\right] &= Am_3m_4\left[\frac{\xi_{3,4}}{\xi_{CAL}^3} - \frac{1}{\xi_{3,4}^2}\right].
\end{aligned}
$$

From the first equation in Eq. 3.29 and the fact that $\xi_{1,3} = \xi_{2,4}$, we have that $m_1m_3 = m_2m_4$. Using a similar argument with the other two expressions leads to

$$m_1m_3 = m_2m_4, \quad m_1m_4 = m_2m_3, \quad m_1m_2 = m_3m_4.$$

It now follows from simple algebra that

$$m_1 = m_2 = m_3 = m_4.$$

The square configuration clearly is a central configuration: it introduces additional equations of the $\frac{\partial g}{\partial \xi_{1,2}} = \frac{\partial g}{\partial \xi_{1,4}}$ form from which the conclusion follows. The real task is to show that it is the only central configuration satisfying these conditions.

To do so, turn to the central configuration equation, Eq. 2.8, which requires each particle's position vector to line up with the particle's acceleration vector. In particular, each of the force vectors $\{\frac{\partial U}{\partial \mathbf{r}_j}\}_{j=1}^4$ must pass through the center of mass, which is the center of the circle. If the configuration is not a square, then, without loss of generality, assume the configuration is of the Fig. 3.11 type and that the common mass value is unity.

In the equation for the force vector

$$\frac{\partial U}{\partial \mathbf{r}_1} = \sum_{j=2}^{3} \frac{\mathbf{r}_j - \mathbf{r}_1}{r_{i,j}^3},$$

the direction $\mathbf{r}_3 - \mathbf{r}_1$ passes through the center, so it contributes to our goal of showing that this is the direction of the force $\frac{\partial U}{\partial \mathbf{r}_1}$. The real concern is to examine what happens in the direction orthogonal to $\mathbf{r}_3 - \mathbf{r}_1$. While $\mathbf{r}_4 - \mathbf{r}_1$ and $\mathbf{r}_2 - \mathbf{r}_1$ are in opposite directions and have the same magnitude in this orthogonal direction, particle 4 is closer to particle 1 than is particle

2. Consequently, because $r_{1,4}^{-1} > r_{1,2}^{-1}$, the orthogonal component from $\frac{\mathbf{r_4}-\mathbf{r_1}}{r_{1,4}^3}$ is greater than that from $\frac{\mathbf{r_2}-\mathbf{r_1}}{r_{1,2}^3}$. Therefore, rather than pointing toward the center of mass, $\frac{\partial U}{\partial \mathbf{r_1}}$ points slightly below. In other words, if this configuration is not a square, it cannot be a central configuration. $\square$

Other results about coplanar central configurations can be obtained in a similar manner.

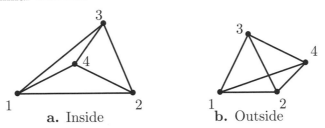

a. Inside   **b.** Outside

Fig. 3.12. Finding constraints

3.6   New kinds of constraints

Eventually we need to use actual ∇g_j constraints where g_j defines either a degenerate tetrahedron or a degenerate pentahedron, so let's do so now. The standard approach is to set the equation for the volume of a tetrahedron equal to zero. But there are other possibilities that, while not in the literature, are surprisingly varied and easy to use. I had intended to introduce these new constraints and then use them to derive new results, but this was not feasible.[18] But I will at least introduce these constraints. What makes them a delight is that some choices involve expressions first learned long ago in high school. I illustrate with the degenerate tetrahedrons of Fig. 3.12.

Immediate choices to describe these degenerate tetrahedrons are:

1. Volume: Set the volume of a tetrahedron equal to zero.

2. Area: If $A_{i,j,k}$ is the area of the $\{i, j, k\}$ triangle, Fig. 3.12a is a degenerate tetrahedron iff

$$A_{1,2,4} + A_{2,3,4} + A_{1,3,4} - A_{1,2,3} = 0. \tag{3.30}$$

A necessary condition for Fig. 3.12b to be a degenerate tetrahedron is

$$A_{1,2,4} + A_{1,3,4} - [A_{2,3,4} + A_{1,2,3}] = 0, \tag{3.31}$$

but this is not a sufficient condition. For instance, Eq. 3.31 is satisfied for an equilateral tetrahedron. To avoid this limitation, one could use (but I do not) that $A_{1,2,3} + A_{2,3,4}$ equals the area of the quadruple defined by the four vertices.

3. Angles: If $\theta_{i,j,k}$ is the angle formed by vertices $\{i, j, k\}$, Fig. 3.12a is a degenerate tetrahedron iff

$$2\pi - [\theta_{1,4,2} + \theta_{2,4,3} + \theta_{3,4,1}] = 0. \tag{3.32}$$

For Fig. 3.12b, we could use

$$2\pi - [\theta_{1,2,4} + \theta_{2,4,3} + \theta_{4,3,1} + \theta_{3,1,2}] = 0. \tag{3.33}$$

An alternative choice that holds for both Figs. 2.14a, b is

$$\theta_{4,1,2} + \theta_{4,1,3} - \theta_{2,1,3} = 0. \tag{3.34}$$

Using volume

The Cayley-Menger determinant for the volume, V, of the tetrahedron is

$$288V^2 = \begin{vmatrix} 0 & 1 & 1 & 1 & 1 \\ 1 & 0 & \zeta_{1,2}^2 & \zeta_{1,3}^2 & \zeta_{1,4}^2 \\ 1 & \xi_{2,1}^2 & 0 & \xi_{2,3}^2 & \xi_{2,4}^2 \\ 1 & \xi_{3,1}^2 & \xi_{3,2}^2 & 0 & \xi_{3,4}^2 \\ 1 & \xi_{4,1}^2 & \xi_{4,2}^2 & \xi_{4,3}^2 & 0 \end{vmatrix} \tag{3.35}$$

Similarly, a power of the volume of the pentahedron is

$$\begin{vmatrix} 0 & 1 & 1 & 1 & 1 & 1 \\ 1 & 0 & \xi_{1,2}^2 & \xi_{1,3}^2 & \xi_{1,4}^2 & \xi_{1,5}^2 \\ 1 & \xi_{2,1}^2 & 0 & \xi_{2,3}^2 & \xi_{2,4}^2 & \xi_{2,5}^2 \\ 1 & \xi_{3,1}^2 & \xi_{3,2}^2 & 0 & \xi_{3,4}^2 & \xi_{3,5}^2 \\ 1 & \xi_{4,1}^2 & \xi_{4,2}^2 & \xi_{4,3}^2 & 0 & \xi_{4,5}^2 \\ 1 & \xi_{5,1}^2 & \xi_{5,2}^2 & \xi_{5,3}^2 & \xi_{5,4}^2 & 0 \end{vmatrix} \tag{3.36}$$

By setting the Eq. 3.35 determinant equal to zero, the constraint of a degenerate tetrahedron is described in terms of leg lengths of Fig. 3.12. I do not pursue this approach, introduced by Dziobek [18], because other approaches seem to be more intuitive.

Using area

To find an expression for Eq. 3.30, recall from Heron's formula learned back in high school that the area of a triangle with leg lengths a, b, c, where $s = \frac{1}{2}(a + b + c)$, is

$$
\begin{aligned}
A &= \sqrt{s(s-a)(s-b)(s-c)} \\
&= \tfrac{1}{4}\sqrt{2(a^2b^2 + a^2c^2 + b^2c^2) - (a^4 + b^4 + c^4)}.
\end{aligned}
$$

Consequently, by symmetry and the law of cosines, all partials have the form

$$
\frac{\partial A}{\partial a} = \frac{1}{8A}a(b^2 + c^2 - a^2) = \frac{1}{8A}a(2bc\cos(\theta))
$$

where θ is the angle between the b and c legs. As $A = \frac{1}{2}bc\sin(\theta)$, the constraint becomes

$$
\frac{\partial A}{\partial a} = \frac{1}{2}a\cot(\theta). \tag{3.37}
$$

Ignoring the $\frac{1}{2}$ multiple (i.e., absorb it in the Lagrange multiplier), it follows from Eqs. 3.37 and 3.30 that the ∇g components for edge legs involve the opposite angles from the exterior and an interior triangle to take the form

$$
\frac{\partial g}{\partial \xi_{1,2}} = \xi_{1,2}[\cot(\theta_{1,4,2}) - \cot(\theta_{1,3,2})], \tag{3.38}
$$

while those for an interior leg involve opposite angles from two interior triangles to assume the form

$$
\frac{\partial g}{\partial \xi_{1,4}} = \xi_{1,4}[\cot(\theta_{1,3,4}) + \cot(\theta_{1,2,4})]. \tag{3.39}
$$

For a clean interpretation of these constraints, absorb all common terms into the Lagrange multiplier. It follows from Eq. 3.17 that

$$
m_j m_k \Big[\frac{1}{\xi_{CAL}^3} - \frac{1}{\xi_{j,k}^3}\Big] = \lambda \frac{1}{\xi_{j,k}} \frac{\partial g}{\partial \xi_{j,k}},
$$

or, using the fact $\lambda < 0$,

$$
\begin{aligned}
\xi_{1,2}^{-3} &= \xi_{CAL}^{-3} + \frac{|\lambda|}{m_1 m_2}[\cot(\theta_{1,4,2}) - \cot(\theta_{1,3,2})], \\
\xi_{1,4}^{-3} &= \xi_{CAL}^{-3} + \frac{|\lambda|}{m_1 m_4}[\cot(\theta_{1,3,4}) + \cot(\theta_{1,2,4})].
\end{aligned} \tag{3.40}
$$

In other words, the cotangent terms describe the necessary modifications to the ξ_{CAL}^{-3} value to obtain $\xi_{j,k}^{-3}$.

For Fig. 3.12b, the constraint structure for the exterior legs is the same as the first of Eq. 3.40 where attention must be paid to the order of the cotangent terms. For instance, since $\xi_{1,2}$ involves $A_{1,2,4}$ and $A_{1,2,3}$, the Eq. 3.31 ordering requires a $\cot(\theta_{1,2,4}) - \cot(\theta_{2,3,4})$ order, while the $\xi_{3,4}$ term requires a $\cot(\theta_{1,3,4}) - \cot(\theta_{2,3,4})$ order. The constraints for the two diagonals assume the form of the second of Eq. 3.40 with a sign difference determined, again, by Eq. 3.31. Namely, the $\xi_{2,3}$ term has a -1 multiple.

The following illustrates the ease of use of these constraints.

Theorem 3.14 *For Fig. 3.12a,b, let $m_2 = m_3$, and assume that a central configuration has $\xi_{1,2} = \xi_{1,3}$. Particle 4 must be on the angle bisector of $\theta_{2,1,3}$: consequently, $\theta_{2,1,4} = \theta_{3,1,4}$ and $\xi_{2,4} = \xi_{3,4}$.*

According to this theorem, if $m_1 = m_2$, then Figs. 2.14a, b (drawn so that $\xi_{1,2} = \xi_{13}$) do not have the required symmetry to be central configurations.

Proof: According to Eq. 3.40, the leg length assumption requires

$$\frac{\lambda}{m_1 m_2}[\cot(\theta_{1,4,2}) - \cot(\theta_{1,3,2})] = \frac{\lambda}{m_1 m_3}[\cot(\theta_{1,4,3}) - \cot(\theta_{1,2,3})].$$

But, triangle $\{1, 2, 3\}$ is isosceles where $\theta_{1,2,3} = \theta_{1,3,2}$. Along with the mass assumption, which requires $m_1 m_2 = m_1 m_3$, it follows that $\theta_{1,4,2} = \theta_{1,4,3}$. The rest of the proof follows from simple geometry. $\square$

Notice, Thm. 3.14 holds for any central configuration of this type that exists for a mixture of positive and negative masses. Another setting is with equal masses. In this manner we capture, in a simple manner, an aspect of Albouy's result [1] that was described earlier (page 46).

Corollary 3.3 *If $m_1 = m_2 = m_3$, then the only four-body central configuration where particles 1, 2, 3 form an equilateral triangle is where the fourth particle is at the center of the equilateral triangle.*

This result, which is true for all choices (positive or negative) of m_4, means that the fourth particle cannot be outside of the triangle. The proof is trivial; the fourth particle must be on the angle bisectors of all three angles.

Angles

Another constraint uses Fig. 3.6b where $\theta_{ik,ij}$ is the angle in triangle i, j, k defined by the legs $\xi_{i,k}$ and $\xi_{i,j}$. This means that $\theta_{12,14} + \theta_{14,13} - \theta_{12,13} \geq 0$ where equality occurs iff the configuration is degenerate and coplanar. If this expression ever is negative, then it corresponds to leg lengths that never allow the configuration to be constructed. By use of the common law of cosines this constraint becomes

$$Arccos(\frac{\xi_{1,4}^2+\xi_{1,2}^2-\xi_{2,4}^2}{2\xi_{1,2}\xi_{1,4}})+ \quad Arccos(\frac{\xi_{1,4}^2+\xi_{1,3}^2-\xi_{3,4}^2}{2\xi_{1,3}\xi_{1,4}}) \\ -Arccos(\frac{\xi_{1,3}^2+\xi_{1,2}^2-\xi_{2,3}^2}{2\xi_{1,2}\xi_{1,3}}) = 0 \tag{3.41}$$

For Fig. 14d, the constraint becomes

$$Arccos(\frac{\xi_{1,4}^2+\xi_{1,2}^2-\xi_{2,4}^2}{2\xi_{1,2}\xi_{1,4}})+ \quad Arccos(\frac{\xi_{1,2}^2+\xi_{1,3}^2-\xi_{2,3}^2}{2\xi_{1,2}\xi_{2,3}}) \\ -Arccos(\frac{\xi_{1,3}^2+\xi_{1,4}^2-\xi_{1,4}^2}{2\xi_{1,4}\xi_{1,3}}) = 0 \tag{3.42}$$

These constraints will be used elsewhere to develop properties of central configurations.

3.7   Rings of Saturn

We now return to the promised discussion of problems related to the rings of Saturn. To do so, we need to introduce a couple more central configurations.

Start with $N \geq 2$ particles, with equal mass, and place them on the vertices of an N-gon where N is an even integer. To determine the force on m_1, pass the line from m_1 passing through the center of mass and meeting another mass on the other side. The $(N - 2)/2$ particles on each side of this dividing line can be categorized into $(N - 2)/2$ pairs consisting of particles from each side of the dividing line that are equal distance from the starting particle. In Fig. 3.13a, for example, one pair consists of $\{m_2, m_8\}$ while another one is $\{m_3, m_7\}$.

To determine the force on m_1, notice from Fig. 3.13a that, by symmetry, each particle in a pair has an equal but opposite pull on m_1 in the y- direction: this direction, given by the dotted lines, is orthogonal to the dividing line. Therefore, the sum of this component is zero. Both particles, however, have an equal pull on m_1 toward the center of mass, so the sum is directed toward the center of mass. As this is true of all pairs, the sum of these forces is toward the center of mass. The remaining particle is diametrically opposite of m_1 (in Fig. 3.13, this is m_5), so its gravitational attraction on m_1 also is toward the center of mass. Thus

$$\mathbf{r}_1'' = \lambda \mathbf{r}_1.$$

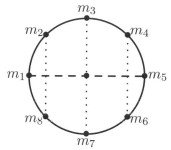

a. Central configuration

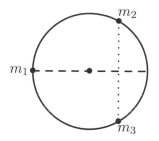

b. Stability?

Fig. 3.13. Central configurations and Saturn

By symmetry, this argument holds for all particles with the same λ. Thus, for this configuration there is a common value of λ so that

$$\lambda \mathbf{r}_j = \mathbf{r}_j'' = \frac{1}{m_j} \frac{\partial U}{\partial \mathbf{r}_j}$$

for each j: the configuration is a central configuration.

As depicted in Fig. 3.13b, the same argument holds if N is an odd integer. (The only difference is that there is no particle diametrically opposite of m_1.) Also notice that we could place a particle, with any desired mass, at the center of mass. Because this centrally located particle pulls equally on each of the other particles, and because it is located at the center of mass (where $\mathbf{r}_{N+1} = \mathbf{0}$), we again have a central configuration. It is this last configuration where the mass of the central particle is chosen to be very large (representing Saturn) and the common masses for the other particles are chosen to be very small (representing a particle in the ring) that Maxwell used to study the rings of Saturn.

3.7.1 Stability

What did Maxwell do? As we know from Chap. 2, it is possible to take any planar central configuration, give it the appropriate velocity, and the configuration will rotate in a circle forever. OK, but is this solution stable? To be more specific, suppose one particle is not precisely where it is supposed to be and it has a velocity that is not quite what we want. Will the solution remain similar to a rotating ring?

To develop intuition, consider Fig. 3.13b where the three ring particles define an equilateral triangle. Now suppose m_1 is slightly above the dashed line. In this situation, m_1 is attracted more toward m_2 than the center of mass, and m_2 feels a similar attraction to m_1. Moreover, there is nothing to balance m_2's affection for m_1 because m_3 is far away. Consequently, we must anticipate that even with a large center mass, this attraction will cause the system to lose its symmetry and stability. Admittedly, it might take a long time for this to happen, but with stability considerations, we have all of eternity to wait for something to go wrong.

The same story does not hold for Fig. 3.13a with the eight ring-bodies. Here m_2 does want to move toward m_1 but the proximity of m_3 provides a retarding effect: it pulls m_2 back in line. This suggests that with enough ring-bodies and with a large enough central force, we should expect some sort of stability to arise.

While we might expect stability, establishing stability for N-body settings is so difficult that it is considered a major unresolved problem. Rather than addressing the stability of the actual problem, we consider the stability of a linearized version of the actual problem. If this linear problem is stable, then we claim that the system is linearly stable.

This is what Maxwell did. He developed a nice analysis showing that if the particle at the center mass (Saturn) has a large enough mass value, and if there are enough particles, then the system is linearly stable.

But, what does it mean to have enough particles? Rick Moeckel [48] addressed this issue by introducing an even weaker version of stability, called spectral stability. To explain, when the linearized version of an equation of motion (about a solution) is used, we have the expression

$$\mathbf{y}' = A\mathbf{y}$$

where A is a square matrix. The solution is linearly stable if the eigenvalues of A are zero, have negative real parts, or are pure imaginaries, *and* if the diagonalized version of A does not have off diagonal terms that require solutions of the $t\cos(\omega t)$ form. Spectrally stable, on the other hand, ignores the matrix structure by just examining the emerging eigenvalues. In this manner, he was able to show that spectral stability occurs for a sufficiently small ring mass iff $N \geq 7$.

Let me see: $N \geq 7$? Seven is the first value where the length between adjacent ring-bodies must be smaller than the distance from the center of mass to the ring. That is, these adjacent distances must be smaller than ξ_{CAL}. Coincidence? Maybe. But I am willing to bet (figuratively) there is a connection. I leave this to others to explore.

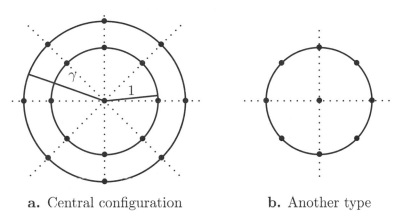

a. Central configuration   **b.** Another type

Fig. 3.14. More rings

3.7.2   More rings

While this all sounds nice, the reality is that Saturn has more than one ring. How can we examine more realistic settings? The first step is to have something to work with; that is, we need to have a central configuration with more rings.[19] My basic idea is captured by Fig. 3.14a.

In Fig. 3.14a, there are two circles with N-bodies equally spaced. Let the inner ring have a fixed radius of unity where all masses have value unity; the outer ring has a radius of $\gamma > 1$ and all masses have value m. The same symmetry argument used earlier shows that each particle in the inner and outer ring satisfies, respectively, the equations

$$\lambda_I \mathbf{r}_j = \frac{\partial U}{\partial \mathbf{r}_j} \quad \lambda_O \mathbf{r}_j = \frac{1}{m}\frac{\partial U}{\partial \mathbf{r}_j} \tag{3.43}$$

where λ_I and λ_O represent, respectively, the common scalar for the particles on the inner and the outer rings. If we can find a value of γ allowing $\lambda_I = \lambda_O$, we have established a central configuration.

To accomplish this goal, an approach is to find how the λ_I, λ_O values change as γ changes. What simplifies our analysis is that for $\gamma \in (1, \infty)$, both terms are continuous functions of γ. But as $\gamma \to 1$, the particles in the inner and outer rings approach each other. As this forces the distances between pairs to approach zero, the particles in the outer ring experience

---

[19] In late 1980's I described the following approach to Edward Slaminka, who was visiting Northwestern, and Kevin Woerner, who was my graduate student, to encourage them to disprove a conjecture by Palmore about central configurations. They succeeded, and as part of their paper [101] they reported and slightly extended my analysis.

an infinite attraction toward the particles in the inner ring. This attraction is directed inwards, which is opposite the position vector of particles on the outer ring, so $\lambda_O \to -\infty$. Each particle in the inner ring also experiences a strong attraction toward its nearest neighbor in the outer ring. But this attraction is being pulled "outwards," which means that the attraction is in the same direction as the position vector. Consequently $\lambda_I \to \infty$.

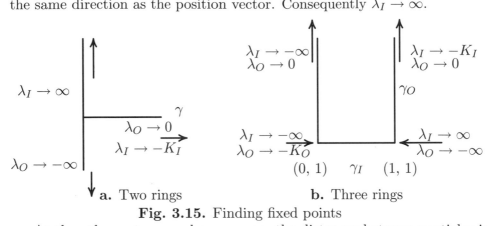

a. Two rings   **b.** Three rings

Fig. 3.15. Finding fixed points

At the other extreme, when $\gamma \to \infty$, the distances between particles in the outer ring approach infinity, so the terms in U, or its partials, involving distances from outer particles to anything else go to zero. Thus, for particles on the outer ring, $\frac{\partial U}{\partial r_j} \to 0$ as $\gamma \to \infty$. But the r_j distances are becoming large, so for $\lambda_O r_j$ to be small, it must be that $\lambda_O \to 0$ as $\gamma \to \infty$.

Now consider λ_I as $\gamma \to \infty$. Here the particles on the inner ring keep a fixed distance from one another, so for these particles $\frac{\partial U}{\partial r_j}$ approaches a fixed force attracting the particles to the center of mass. Thus $\lambda_I \to -K$, where K is some positive value, as $\gamma \to \infty$. By graphing these two curves over $(1, \infty)$ (this is indicated in Fig. 3.15a), it is clear they must intersect, Such an intersection defines a common value for $\lambda_I = \lambda_O$; it establishes a γ value for which there is a central configuration.

Three, or four, or ..., rings

We could carry out this same analysis out for three rings of radii $1 < \gamma_1 < \gamma_2$. Here there are three λ values: $\lambda_I, \lambda_M, \lambda_O$ for the inner, middle, and outer rings. The only change in the analysis is that the λ values are functions of two variables over the region $1 < \gamma_1 < \gamma_2 < \infty$: the λ's define a surface over the indicated region. Again, by examining what happens to each λ near the boundaries of this region, we can establish that there is a central configuration with three, or four, or ... rings.

What about the stability of these rings? While I would love to discuss the actual stability, I would settle for information about the linear or spectral stability. While never examined, this question seem doable.

Still another class

To create still another class of central configurations to serve as models for rings of Saturn, return to Fig. 3.14b. We already have established that the configuration, as it stands, is a central configuration. But what if we take every other particle—say the ones on the dotted lines—and split it into two particles: the inner portion has mass $m_I > 0$ while the common mass value for the outer portion is $1 - m_I = m_O > 0$. Now, move the portions with m_I masses inwards with radius $0 < \gamma_I < 1$, and the other portion outwards with radius $1 < \gamma_O < \infty$ to generate another three-ring operation. The same symmetry arguments show that the particles on each of the three rings satisfy the equations

$$\lambda_k \mathbf{r}_j = \frac{1}{m_j} \frac{\partial U}{\partial \mathbf{r}_j},$$

where m_j equals unity, m_I, and m_O, respectively, for particles on the middle ring of fixed radius unity, the inner ring, and the outer ring. The lambda value for each ring is $\lambda_M, \lambda_I, \lambda_O$ for the middle, inner, and outer rings. Using essentially the same analysis as above, we can show that there is an appropriate spacing that gives $\lambda_M = \lambda_I = \lambda_O$, which defines still another central configuration.

The argument is depicted in Fig. 3.15b where each λ now defines a surface over the rectangle defined by γ_I and γ_O. Again, the approach is to examine the boundary values of these surfaces: this is indicated in Fig. 3.15b. As indicated on the left edge, the surface λ_I approaches $-\infty$ as the particles approach a collision, or when $\gamma_I \to 0$. When this happens, it separates the rings so λ_O approaches a negative constant as $\gamma_O \to 1$, or zero as $\gamma_O \to \infty$. Similar arguments handle the other edges.

The λ_M surface does not have extreme behavior on the boundaries. On the bottom edge, where $\gamma_O = 1$, the λ_M surface ranges from what it would be with a larger mass in the center (with $\gamma_I = 0$) to the value it would have if the particles were not carved up (at $\gamma_I = \gamma_O = 1$). A similar argument holds on the edges. On the other hand, the extreme behavior of λ_I requires the surfaces λ_M and λ_I to intersect and define a curve starting on the bottom edge. This curve terminates on the right-edge (meaning that $K_I < -K_M$) or it continues up the long rectangle to infinity (meaning that as $\gamma_O \to \infty$, the λ_M value is less than $-K_I$). It remains to show that this curve intersects the surface defined by λ_O.

That the curve intersects the λ_O surface follows from the boundary values of λ_O. For instance, if $-K_O > \lambda_M$ for some choice of γ_I on the lower boundary, we have our intersection. If this is not the case, then because $\lambda_O \to 0$ as $\gamma_O \to \infty$, but λ_M always is negative and bounded away from zero, there must be an intersection in the interior of the rectangle.

With the new central configurations, the question is to examine the linear and spectral stability of these settings. If N has a sufficiently large value and if the masses are sufficiently small, do we still retain stability? None of these systems have ever been examined in this manner.

3.7.3   Saturn, and some problems

Chapter 1 was directed, in part, to provide background for the question about the dynamics of the F-ring of Saturn. The material in Chaps. 2 and 3 are brought together here with a question about the stability of the rings of Saturn. While easily stated, these problems have yet to be resolved.

There is another class of problems. The above arguments finding central configurations by the intersection of curves and surfaces is based on symmetry configurations. There are many other symmetry settings to consider; e.g., what happens by starting with kN bodies equally placed on a circle (or sphere), and then moving certain ones inwards and others outwards at different rates? We should be able to get even more central configurations. The problem is to establish that the different λ_j values must equal.

The argument establishing the equality of the different λ_j values—the existence of the central configurations—describes the λ_j behavior on the boundaries. Namely, these arguments are just disguised versions of standard fixed point, or index arguments. This suggests that by using fixed point arguments and symmetry configurations it should be possible to obtain a general result that characterizes this huge class of central configurations.

Chapter 4

Collisions – both real and imaginary

The rest of this book stresses a particular topic where central configurations play a major role: collisions. To provide a more inclusive discussion, I discuss singularities, collisions, and near collisions.

The obvious first goal is to understand what are singularities, and why we should care about them. Namely, we need to determine how singularities affect the behavior and our knowledge of the N-body problem. This is the theme of the current chapter. To motivate and introduce some of the themes, I first describe where and why some of these issues arise.

A well known story about the N-body problem involves H. Poincaré's incorrect solution of the restricted three-body problem: he thought that he solved it, but he did not.[1] Adding drama to the tale is how it involved a large prize, the King of Sweden, and how Poincaré's alternative "prize paper," which essentially described why he did not solve the three-body problem, introduced the foundations of what we now call "chaos." What is not so well known is that a couple of decades later, in an amazing piece of work, the Finnish mathematician–astronomer Karl Sundman[2] solved the

---

[1] This story, which has been well-known among experts in celestial mechanics through Winter's book [112] and other sources, has been made available for others through the expository book by Diacu and Holmes [16], the comments by Peterson [57] in his general audience book, and the careful historical description by Barrow-Green [4] in her book.

[2] The literature describes Karl Fritiof Sundman (1873-1949) as a "Finnish mathematician." I concur: the quality and thrust of his research is that of an excellent mathematician. The history of his career, however, imposes a minor inconvenience to this description: Sundman's position was as a Professor of Astronomy and as the Head of the Helsinki Observatory. He started his career at one of the leading nineteenth century observatories:

full three body problem according to the standards of his time. There is a natural question: if Sundman "solved" the three-body problem, why is his solution not that well known?

To explain, up until the early twentieth century, a preferred solution approach for the N-body problem was to find a convergent power series expansion. By accepting and adopting this approach, the role of singularities becomes clear: as even students in a calculus course quickly learn, the radius of convergence for these series is determined by the distance to the nearest singularity.[3] To obtain a solution that exists for all time, Sundman cleverly attacked and partially conquered the troubling issue of "singularities." But any reader with a sufficiently vivid imagination can envision how after he finished his impressive work, in a search for revenge, the mathematical singularities of the N-body problem waged a counterattack on Sundman— and they partially won. What the singularities of the system did was to impose sufficiently severe constraints on the Sundman's series solution to make them totally useless from a practical perspective: the series converge much too slowly to be of any value.

While Sundman's series are of no practical use, almost a century later his contributions toward our understanding of collisions and the N-body problem remain valued tools in our analysis of celestial mechanics. Indeed, a brief review of Sundman's story identifies other N-body structures and leads to some nice research questions. They are discussed next.

4.1   One body problem

To develop insight into the behavior of collisions, start with the one-body, or central force problem

$$\mathbf{r}'' = -\frac{\mathbf{r}}{r^3}. \tag{4.1}$$

From standard existence theorems of differential equations, we know that the solution exists and can be extended as long as $r \neq 0$. Presumably, the negative situation that allows $r \to 0$ must correspond to a collision. Indeed,

---

the Pulkovo Observatory of the Russian Academy of Sciences that is located near Saint-Petersburg. (Finland did not win its independence from Russia until 1917.) Most surely due to Sundman's influence, during the first part of the twentieth century, the research emphasis at the Helsinki Observatory focused on the theoretical studies of celestial mechanics. In recognition of his important research contributions, Sundman became a fellow of learned societies in several countries and twice he received the Pontécoulant Prize.

[3]The history of this classical result is intimately related to the N-body problem. See Section 4.3.7.

without collisions, the solutions of this equation are ellipses, parabolas, or hyperbolas.

But what happens near or at a collision? Is it possible to define the motion through $r = 0$? These are the kinds of questions Sundman [106, 107] posed and resolved in order to develop his theory. Rather than describing Sundman's more complicated argument, a simpler approach describing the passage through a collision, which was developed about a decade later in 1920 by the Italian geometer Levi-Civita[4] [38], is the basis of the following.

To provide motivation, start with a family of orbits that suffer a collision in the limit. Such a family of elliptic orbits

$$r = \frac{a}{1 - \epsilon \cos \theta} \tag{4.2}$$

is depicted in Fig. 4.1 where, when the eccentricity $\epsilon \rightarrow 1$, the ellipses become narrower and narrower until, in the limit, the collinear solution would hit the body (the bullet), and, presumably, pass through 360° to rebound backwards.

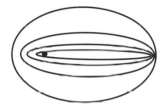

Fig. 4.1. Approaching a collision orbit

According to Eq. 4.2 and the energy integral

$$\frac{1}{2}v^2 = \frac{1}{r} + h, \tag{4.3}$$

---

[4]Tullio Levi-Civita (1873-1941) was an eminent mathematician who, in addition to celestial mechanics, worked in a variety of areas including partial differential equations, hydrodynamics, relativity, quantum theory, and even some analytic number theory. He is best known for his work with his advisor, Ricci-Curbastro, his work on the concept of parallelism in curved spaces, and the development of tensor analysis. These advances, of course, proved fundamental for A. Einstein's work on relativity and many other topics. During the years leading up to World War II, Levi-Civita, along with many other Italian mathematicians and scientists, actively opposed Fascism. Because of the anti-Semitic laws (he was of Jewish descent), in 1938 Levi-Civita was dismissed from his academic position in Rome and expelled from all Italian scientific societies. His health declined, and he died a couple of years later (December 29, 1941) after a stroke. Among the high honors recognizing his accomplishments is, appropriately, Crater Levi-Civita on the moon.

when r reaches its smallest value the particle has a "close approach " where the particle rapidly spins around the central body. Therefore, in the limit at a collision we should anticipate that the colliding particle makes a 2π angular change—this suggests that perhaps the particle hits and then rebounds from the central force in a manner resembling a perfectly elastic rebound. This reasoning suggests that in order to remove this collision singularity from the equations of motion, the first task is to find a way to replace the sharp 360° turn by "straightening out" the orbit.

4.1.1   Levi-Civita's approach

To see how Levi-Civita's approach straightens out the orbit, express $\mathbf{r} = (x, y)$ as $z = x + iy$ and the equations of motion, $\mathbf{r}'' = -\frac{\mathbf{r}}{r^3}$, in complex variable form as

$$z'' = -\frac{z}{r^3}. \tag{4.4}$$

The collision, which occurs at $r = 0$, is precisely where the change of variables must convert the abrupt 2π polar angle change into a form that keeps the motion is on a straight line. The objective is to retain only half of the angular change at the collision, so a natural choice of new variables is the square root; i.e., consider the coordinate change $w = u_1 + iu_2 = z^{1/2}$, or

$$w^2 = z. \tag{4.5}$$

As shown next, the equations of motion for w, when accompanied with the change of the independent variable $ds = dt/r(t)$ that was introduced by Sundman, go beyond creating a system that is well defined at $w = 0$ to have one that assumes the particularly simple and powerful (by being linear) form

$$\mathbf{u}'' + a\mathbf{u} = \mathbf{0} \tag{4.6}$$

where a is a positive constant and $\mathbf{u} = (u_1, u_2)$. In other words, *the nonlinear Newtonian equations Eq. 4.1 are converted into the linear equations for a harmonic oscillator.*

To derive Eq. 4.6, notice from calculus that the change of the independent variable defines the operator

$$\frac{d^2}{dt^2} = \frac{r\frac{d^2}{ds^2} - \frac{dr}{ds}\frac{d}{ds}}{r^3}.$$

The r^3 denominator plays an important role because it cancels the r^3 denominator in the Newtonian force. Indeed, applying this operator to the equations of motion (Eq. 4.4) immediately leads to the expression

$$rz'' - r'z' + z = 0$$

where the primes now indicate differentiation with respect to s. Similarly, because $\frac{d\mathbf{r}}{dt} = \frac{1}{r}\frac{d\mathbf{r}}{ds}$, the energy integral $\frac{1}{2}v^2 = \frac{1}{r} + h$ becomes

$$(z')^2 = 2(r + hr^2).$$

After introducing the new dependent variables u_1 and u_2, terms cancel by replacing r and r' with their representation in terms of w and its conjugate. The equations of motion and the energy integral now are

$$2w'' - \frac{w}{r}(2|w'|^2 - 1) = 0, \quad 2|w'|^2 = 1 + hr. \tag{4.7}$$

In other words, by using the energy integral and letting $a = -h/2$ (elliptic motion requires $h < 0$ so $a > 0$), not only does Eq. 4.6 emerge, but we now know that the transformed equations are restricted to a fixed energy surface.

Advantages

Beyond being able to analyze the behavior of orbits near a binary collision, an interesting practical advantage of Eq. 4.6 is that the solutions of this linear equation are stable. This means that even with small perturbations, the solutions preserve much of the original behavior. A particularly important consequence is that the numerical solutions retain the properties of the actual solution.

What makes this pragmatic contribution for numerical solutions so useful is that a similar stability comment fails to hold for numerical solutions of Eq. 4.1. Instead, with Eq. 4.1, a small numerical error can force the numerical solution onto a different energy surface that, in turn, alters the frequency of the motion. Then, as true for two pendulums with close but different frequencies, the true and the computed solutions eventually can even be at opposite ends of the orbit. In contrast and as exhibited by the derivation, the transformed Eq. 4.6 keeps all solutions on the same energy level. Consequently the "regularized" system Eq. 4.6 is much preferred for numerical studies.

4.1.2 Kustaanheimo and Stiefel's approach

Several important dynamical systems, such as the Earth satellite problem, are perturbed forms of Eq. 4.1. (The perturbations of the equations for the

Earth satellite problem reflect Earth's lack of homogeneity; to appreciate the loss of homogeneity, think of the huge mountains, deep oceans, and the Earth's "middle age bulge" centered around the equator.) Because, in general, these systems cannot be solved analytically, they must be numerically integrated. Consequently, it is natural to question whether perturbed forms of Eq. 4.1 can be converted into a form that inherits at least some of the stability properties of the harmonic oscillator. Moreover, because "real" problems such as the Earth satellite problem reside in a three-dimensional setting, an obvious goal is to convert the three-dimensional version of Eq. 4.1 into the equation for a three-dimensional harmonic oscillator. There are comments in the literature (e.g., page 23 of Stiefel and Scheifel [105]) suggesting that Levi-Civita tried to find such a three-dimensional extension, but he was unsuccessful. I will explain why below.

As the world moved into the Space Age, this issue of finding a harmonic oscillator formulation for the three-dimensional two-body problem added practical importance to its theoretical interest. Consequently, it was natural for Eduard Stiefel (1909-78) to raise this issue during the 1964 Oberwolfach conference on celestial mechanics that he organized. Attending this conference was Paul Kustaanheimo (1924-97), a Finnish mathematician from the University of Helsinki, who had been using spinors and quaternions to analyze problems such as symmetry in the Einstein Field Equations as well as with other mathematical concerns from astronomy and physics.

Kustaanheimo's approach

Quaternions (introduced by William Hamilton in 1843—presumably discovered while he was taking a walk with his wife in Dublin just when crossing the Brougham Bridge now called Broom Bridge) are a natural generalization of complex variables. The evening Kustaanheimo learned about the challenge posed by the Levi-Civita system, he retired to his room to try to mimic the Levi-Civita approach by replacing complex variables with quaternions. He was successful. More specifically, let

$$w = u_1 + iu_2 + ju_3 + ku_4, \quad z = x_1 + ix_2 + jx_3 + kx_4$$

and $\mathbf{u} = (u_1, u_2, u_3, u_4)$, where w and z satisfy the quaterion rules $i \cdot i = j \cdot j = k \cdot k = -1$, $i \cdot j = -j \cdot i = k$, $j \cdot k = -k \cdot j = i$, and $k \cdot i = -i \cdot k = j$. Kustaanheimo [30] used the same $w^2 = z$ change of dependent variables along with Sundman's change of independent variables to convert the $\mathbf{r} \in \mathbb{R}^4$ version of Eq. 4.1 into the harmonic oscillator equations

$$\mathbf{u}'' + a\mathbf{u} = \mathbf{0}, \quad \mathbf{u} \in \mathbb{R}^4. \tag{4.8}$$

There is a complication: the true problem resides in $\mathbb{R}^3$, not $\mathbb{R}^4$. But, as indicated later, this change of variables maps each Eq. 4.8 solution to a fixed plane $x_4 = c$. Thus, by treating x_4 as a dummy variable that can be set equal to zero, the easily obtained solutions of Eq. 4.8 harmonic oscillator equation are mapped to solutions of Eq. 4.1 in the $x_4 = 0$ plane. Moreover, because a point z in the $x_4 = 0$ plane determines the magnitude of the corresponding $\mathbf{u}$, it follows from the dimensional differences that each point in the plane $x_4 = 0$ corresponds to a circle in the $\mathbf{u}$ variables rather than just the two points as true for the Levi-Civita transformation. The important exception is when $\mathbf{r} = \mathbf{u} = \mathbf{0}$ where the circle degenerates into a point.

Now that the quaternion transformation solved this decades old problem (it can be done with spinors too), Kustaanheimo and Stiefel [31] joined forces to use this approach to analyze perturbed versions of Eq. 4.1. Among other conclusions, they showed that their transformation does provide numerical advantage for earth satellite systems. (Extensions are developed in Stiefel and Scheifel [105].)

4.1.3 Topological obstructions and hairy balls

With all of the mathematical talent that had been searching for a way to extend Levi-Civita's approach, an interesting mystery is to explore what frustrated their attempts. For instance, why can't we find a similar relation for $\mathbb{R}^3$ without having to jump a dimension to invoke $\mathbb{R}^4$?

To understand where difficulties arise, express Levi-Civita's $w^2 = z$ change of dependent variables, $x = u_1^2 - u_2^2$, $y = 2u_1u_2$, in the differential form

$$\begin{pmatrix} dx \\ dy \end{pmatrix} = 2 \begin{pmatrix} u_1 & -u_2 \\ u_2 & u_1 \end{pmatrix} \begin{pmatrix} du_1 \\ du_2 \end{pmatrix}. \tag{4.9}$$

The first column of this orthogonal matrix can be viewed as locating a point on the unit circle $u_1^2 + u_2^2 = 1$ while the second column defines a unit tangent vector to the circle at this point $\mathbf{u}$. This description, which uses the two column vectors, identifies the Levi-Civita transformation with an unit vector field on the unit circle S^1 in $\mathbb{R}^2$.

Presumably the appropriate $\mathbb{R}^3$ change of dependent variables creating the harmonic oscillator would be represented by a 3×3 orthogonal matrix of the Eq. 4.9 type where the first column locates a point on S^2, the unit ball in $\mathbb{R}^3$, while the remaining two columns correspond to unit tangent vectors to the sphere that are orthogonal to one another. The problem, of course, is that such a matrix does not exist: if it did, it would be equivalent to a continuous combing of a hairy billiard ball without encountering a "cowlick" where at least one hair stands upright. But as we know from Poincaré and Brouwer, there does not exist a 3×3 orthogonal matrix of this appropriate form: these topological reasons, then, are what proscribe the existence of the desired change of variables.

On the other hand, reaching back to his Ph.D. dissertation and earlier work in algebraic topology,[5] Stiefel was well aware that while it is impossible to comb a hairy S^2, it is possible to comb a hairy S^3 (the unit ball in $\mathbb{R}^4$) in several ways. Stated in another manner, this unit ball admits a smooth frame[6] where if (u_1, u_2, u_3, u_4) identifies a point on S^3, then three mutually orthogonal, unit tangent vectors are

$$(-u_2, u_1, u_4, -u_3), \; (-u_3, -u_4, u_1, u_2), \; (u_4, -u_3, u_2, -u_1).$$

By expressing these four vectors as columns of a 4×4 matrix, much as in Eq. 4.9, the Kustaanheimo-Stiefel transformation of dependent variables is

$$\begin{pmatrix} dx_1 \\ dx_2 \\ dx_3 \\ dx_4 \end{pmatrix} = 2 \begin{pmatrix} u_1 & -u_2 & -u_3 & u_4 \\ u_2 & u_1 & -u_4 & -u_3 \\ u_3 & u_4 & u_1 & u_2 \\ u_4 & -u_3 & u_2 & -u_1 \end{pmatrix} \begin{pmatrix} du_1 \\ du_2 \\ du_3 \\ du_4 \end{pmatrix}. \tag{4.10}$$

As it is easy to check, the last row of this matrix is orthogonal to **u**, and this ensures that $x_4 = 0$. Indeed, as some readers most surely will recognize, all of this is intimately related to the Hopf maps from S^3 to S^2. To complete the intellectual circle, Hopf was Stiefel's Ph.D. adviser.

4.1.4   Sundman's solution of the three-body problem

Let me now return to Sundman's masterful work of finding converging series solution for the three-body problem. Of course, a serious complication

---

[5] An amusing incident that captures Stiefel's eclectic interests and mathematical versatility occurred during an 1978 Oberwolfach meeting when a mathematician leaned over to confide how he was impressed that Stiefel, who was widely known for his contributions to computer science, numerical analysis and celestial mechanics, could incisively comment on an algebraic topological argument that had just been used to describe a particular N-body property. In response, I asked whether he ever heard of the Stiefel-Whitney classes or the Stiefel manifolds. He knew I was setting him up, so he volunteered his guess that that Stiefel was our host's father. As I explained, no, this is the man himself.

[6] When I spent a term with Stiefel at E.T.H. in Zurich, we discussed how this nice behavior holds only for S^1, S^3, and S^7. The Levi-Civita structure uses S^1, the Kustaanheimo-Stiefel approach uses S^3. Are there natural uses of S^7 for celestial mechanics? Only briefly did we think about this and I never returned to the issue—but I wonder.

hindering the construction of a convergent power series solution for the three-body problem are those singularities caused by binary collisions. As it will be shown in Chap. 5, collisions of any kind are improbable. Nevertheless, collisions do exist, and whenever they do occur, they impose restrictions on the radius of convergence for any power series solution. A possible way to avoid these complications would be to restrict attention to those initial conditions with collision-free orbits. The obvious fault with such a program is that we do not know how to characterize these conditions. Indeed, in general, we have no way of knowing whether an initial condition will, or will not, lead to a collision-free trajectory.

As already indicated, Sundman cleverly avoided the problem of binary collisions by converting the equations into a system where a binary collision is *not* a singularity—it merely is another regular point. Because of the central role played by his result, it is worth spending a couple of paragraphs indicating why the Sundman change of the independent variable,

$$ds = \frac{dt}{r(t)}, \tag{4.11}$$

is the "natural" one.

A way to start the discussion is to determine the behavior of colliding particles in the collinear central force problem. If x is the distance between two particles, then the defining equation of motion is

$$x'' = -\frac{1}{x^2}. \tag{4.12}$$

As the right-hand side of this equation always is negative, the solution $x(t)$ is concave down. Consequently, the system must suffer a collision either forwards or backwards in time: e.g., if at time t we have that $x'(t) \leq 0$, then it must be that x' remains negative in the future: the doomed particle is headed for a terminating collision.

To determine the behavior of the collision, multiply both sides of Eq. 4.12 by x' and integrate to obtain the energy integral

$$x'^2 = \frac{2}{x} + 2h \tag{4.13}$$

where h is a constant of integration. Since a collision occurs at time $t = t^*$ if and only if $x \to 0$ as $t \to t^*$, it follows that when approaching a collision, the energy integral defines the asymptotic form $x(x')^2 = 2 + 2hx \sim 2$. In turn, this means that $x^{1/2}x' \sim -\sqrt{2}$, or $x^{3/2} \sim \frac{3}{2}\sqrt{2}(t^* - t)$. Consequently,

$$x(t) \sim A(t^* - t)^{2/3} \quad \text{where } A = (\frac{3}{2}\sqrt{2})^{2/3} \quad \text{as } t \to t^*. \tag{4.14}$$

As shown later in this chapter (and motivated by the arguments of Sect. 2.2.1 (page 36)), if there is a collision of any kind for the general N-body problem, then the colliding particles must approach each other like $(t^* - t)^{2/3}$; namely, this $\frac{2}{3}$ power describes what happens for all possible kinds of collisions.

Even more is known. As also shown later, binary collisions always correspond to algebraic branch points. (Three and higher body collisions can be logarithmic branch points.) This algebraic singularity statement remains true even if several different binary collisions occur simultaneously. In particular, in a neighborhood about the time of collision, the solution is

$$x(t) = \sum_k a_k (t^* - t)^{2k/3}. \tag{4.15}$$

Equation 4.15 means that *a binary collision for the system is an algebraic branch point* where $s = (t^* - t)^{1/3}$ serves as a local uniformizing variable. This change of independent variable converts Eq. 4.15 into the analytic expression $x(s) = \sum_k a_k s^{2k}$. Consequently, as t passes through t^*, the solution races through the sheets of the Reimann surface to emerge as thought the collision corresponds to an exact rebound (as captured by the fact that all exponents of s are even), or a perfectly elastic collision.

We now can provide an intuitive argument why Sundman's change of the independent variable works. According to Eq. 4.15, by substituting $s = (t^* - t)^{1/3}$, or

$$ds = \frac{-dt}{3(t^* - t)^{2/3}}, \tag{4.16}$$

we can obtain an analytic solution that passes through a binary collision. The problem, however, is that we do not know whether a collision will occur with a specified initial condition. Even if, in some clever manner, we could determine that a collision will occur, we have absolutely no idea how to find the precise time value t^* when the binary collision happens.

A substitute approach that cleanly avoids both of these complications is to base the time change on the appropriate power of $r(t)$. After all, a collision occurs if and only if r approaches zero, so the growth properties of r determine whether and when a collision will happen. Since $r(t) \sim A(t^* - t)^{2/3}$, it follows from Eq. 4.16 that the natural choice is $ds = dt/r(t)$: this is Sundman's change of the independent variable. (I dropped the minus sign: it just changes the direction of s so that it corresponds to that of t.)

4.2 Sundman and the three-body problem

With this description of Sundman's treatment of binary collisions, we can describe how he addressed the three-body problem. An outline of his beautiful work is given in this section.

After Sundman eliminated binary collisions from the equations of motion, the only remaining singularity for the transformed three-body problem is a triple collision. Triple collisions can be avoided by appealing to a result proved by Sundman, and already known by Weierstrass:[7] an assertion involving the Eq. 2.3 integral of angular momentum $\sum_j m_j \mathbf{r}_j \times \mathbf{v}_j = \mathbf{c}$. The Weierstrass-Sundman theorem is the following:

Theorem 4.1 *(Weierstrass-Sundman) For the three-body problem, if* $\mathbf{c} \neq \mathbf{0}$, *then the system cannot have a complete collapse; i.e.,* $\mathbf{R}$ *cannot approach zero.*

Restated in mathematical terms, if $N = 3$, it is impossible for a triple collisions to occur off the algebraic variety defined by $\mathbf{c} = \mathbf{0}$. A proof of this theorem with some consequences is the theme of Sect. 4.3.

4.2.1 Complex singularities?

By restricting attention to $\mathbf{c} \neq \mathbf{0}$, Sundman removed all real singularities from the transformed equations for the three-body problem. But difficulties remain because the radius of convergence of a series solution is determined by the nearest singularity independent of whether the singularity is real

---

[7] We normally associate Karl Weierstrass (1815-97), often called the "Father of Analysis," with the development of mathematical rigor and his many analytic contributions, but not celestial mechanics. Perhaps it was our first experience as students struggling with $\epsilon - \delta$ arguments that incorrectly paints the picture of Weierstrass as being a grim, strict, stoic bearded person. Well, he did have a beard, but I find him to be colorful and interesting. For instance, perhaps in rebellion against his overbearing and meddling father, who wanted Weierstrass to become a "bookkeeper" or work in governmental service as a public administrator, Weierstrass thoroughly enjoyed his undergraduate days at the University of Bonn by skipping classes, sword-fighting, beer drinking—and reading mathematics books. His fascinating life, filled with heart-ache, excellence in teaching and exposition, and his mathematical successes that centered around analysis, power series, the "Weierstrass Elliptic Functions," and others, also included his expertise and contributions to the N-body problem. Indeed, it was Weierstrass who posed to his former student, Mittag-Leffler, the problem mentioned in the introductory paragraphs of this chapter that Poincaré addressed. Even more, Sophia Kovalevsky, who studied with Weierstrass (she could not attend his public lectures because she was a "she," so Weierstrass tutored her privately) and remained close to him, analyzed the rings of Saturn via partial differential equations. For added information about Weierstrass, see Mittag-Leffler [46].

or complex-valued. Thus Sundman now had to confront the possibility of complex-valued collisions. This raises a new issue: is it possible to have imaginary collisions of the particles that occur at imaginary values of time?

To show that such imaginary collisions can occur, consider the elliptical solutions of the central force problem given by Eq. 4.1. The solution for $r(t) = (\mathbf{r}, \mathbf{r})^{1/2}$ is defined implicitly through the Kepler equations

$$r(u) = a(1 - \epsilon \cos(u)), \quad t = u - \epsilon \sin(u) \tag{4.17}$$

where ϵ is the eccentricity of the ellipse, a is the length of the semimajor axis, and u is a variable.

Because the elliptical solutions of the two-body problem are well behaved without any possibility of a collision, it is reasonable to wonder whether the associated power series solution converges for all values of time. This is not the case: the limits on the radius of convergence are imposed by the location of complex singularities: they are caused by imaginary collisions that take place at imaginary values of time. By letting $u = u_1 + iu_2$ and $t = t_1 + it_2$, a simple computation, which just involves the expansion of sine and cosine for complex values, proves that there exist imaginary collisions where $r = 0$, and they occur at

$$u_k = 2k\pi \pm i\cosh^{-1}(\frac{1}{\epsilon}), \quad k = 1, 2, \ldots, \quad t_k = u_k - \epsilon \sin(u_k). \tag{4.18}$$

4.2.2   Avoiding complex singularities

The point to be made is that complex singularities do exist, so they can determine the radius of convergence of a power series solution. Notice that the real part of the time of an imaginary collision, where $Re(t_k) = 2k\pi$, is precisely the time on the real time axis when $r(t)$ assumes its minimum value.[8] Moreover, the smaller the value of this minimum (i.e., the larger the value of ϵ), the closer this complex singularity is to the real axis. These comments suggest that if the three-body problem is "bounded" away from a triple collision, then maybe the complex singularities are bounded away from the real axis.

This is precisely what Sundman proved. His first step, which is a significant improvement over the Weierstrass-Sundman Theorem, was to prove that if $\mathbf{c} \neq \mathbf{0}$, then associated with each orbit is a constant $D(\mathbf{c}) > 0$ so that for all t we have that $r_{max}(t) = \max(r_{j,k}(t)) \geq D(\mathbf{c})$. In other words, should

---

[8] Does something like this hold for the N-body problem? That is, does the location of minima of U locate complex singularities? Results of this kind can be obtained.

$\mathbf{c} \neq \mathbf{0}$, then not only are there no triple collisions, but the maximum spacing of particles in the system is strictly bounded away from zero and protected from any involvement in a triple collision. In fact, a stronger assertion can be proved: there is a value D such that if r_{max} ever is less than D, then $r_{max}(t) \to \infty$ as $t \to \infty$.[9] Later, Marchal and Saari [40] extended all of these assertions from the three-body problem to the general N-body problem: one of these general N-body results is derived in Sect. 4.3.

Conformal mapping

As suggested by our intuition developed with Kepler's equations Eq. 4.17, once Sundman found a lower bound on how close all particles can approach each other, he was able to prove that the system has no complex singularities in a strip, where its width depends on the value of $\mathbf{c}$, in the complex plane centered about the real axis. With this conclusion, the rest of Sundman's proof is immediate. All one needs to do is to conformally map this infinite strip to the unit disk. For instance, if the strip is defined by $|\text{Im}(s)| \leq \beta$, then it suffices to use the change of independent variable

$$\sigma = \frac{e^{\frac{\pi s}{2\beta}} - 1}{e^{\frac{\pi s}{2\beta}} + 1}.$$

In this new system, the equations of motion have no singularities in the unit disk, so the resulting power series converges for all $|\sigma| < 1$. Along the real axis of this unit disk, σ corresponds in a 1 1 fashion to all real values of time t. Consequently, Sundman proved the existence of a power series solution for the Newtonian three body problem that converges for all real values of time!

4.2.3 Singularities retaliate

Although Sundman solved this historic problem in a masterful and accepted manner of his era, the unfortunate reality is that his series solution is of minimal to no practical value. This is because the series tends to converge much too slowly and requires too many terms to achieve any interesting degree of accuracy.

The source of this slow convergence is easy to understand in terms of Sundman's two changes of the independent variable. The first change, which eliminated the singularities caused by binary collisions, behaves like

---

[9]I am not sure whether Sundman, or G. D. Birkhoff using Sundman's ideas, proved this result.

$s = (t^* - t)^{\frac{1}{3}}$ near a binary collision (Eqs. 4.11, 4.14). This change of variables significantly "slows down" the dynamics with the accompanying effect that a numerical value of s tends to be much larger than the corresponding numerical value of t. To appreciate what is happening, notice the similarity with what happens in numerical integration where smaller step sizes are used to achieve accuracy with a rapid change in the dynamics. Indeed, with the above change in variables and near a collision, the step sizes would have the $(\Delta s)^3 \approx \Delta t$ relationship, which means that a small Δs corresponds to a miniscule Δt. Accompanying the smaller steps is a significant increase in the number of steps that are needed to reach and represent a s value.

After binary collisions are handled, Sundman exponentially mapped these larger s values to the unit disk in the complex σ-plane. But as starting students in complex variables quickly learn, even fairly small values of s, with potentially even smaller values of t, can be identified with σ values that are very close to the boundary of the unit disk. Whenever this happens, expect an agonizingly slow convergence. This is precisely what happens.

Although Sundman successfully conquered the long standing challenge of "solving" the Newtonian three-body problem, the theoretical properties of his analysis demonstrate that, in general, an universal power series solution of the three-body problem is impractical. In other words, his success in solving an age-old problem successfully killed interest in this line of inquiry. Nevertheless, many of Sundman's results used to achieve his series solution continue to play a significant role in the development of celestial mechanics. More about Sundman and his contributions will be described in Saari [90].

4.3   Generalized Weierstrass-Sundman theorem

It is important to point out that the above results using the $t^{2/3}$ approach, such as the ability to regularize collisions and some of the other collision properties, need not extend to other force laws. For instance, for the inverse p force law, the $t^{2/3}$ asymptotic governing the approach of colliding particles is replaced with a $t^{2/(p+1)}$ leading term. As described earlier, the "elastic collision" behavior occurs because as a collision is approached, the solution races through the sheets of the Reimann surface to emerge along the real axis: it is not difficult to show that this need not happen with other p values. As another place where the force law makes a difference, recall that an important part of the Sundman program was the Weierstrass-Sundman theorem that complete collapse is impossible if $\mathbf{c} \neq \mathbf{0}$: this assertion does not hold for $p \geq 3$.

4.3.1 A simple case–the central force problem

Before proving an extended N-body version of this result, it is worth developing intuition as to why the Weierstrass-Sundman theorem holds. A way to do so is to examine what happens with the central force law where a similar effect already occurs. A simple way to explain is to use the equations of motion for the length r. To derive these equations, by differentiating $r^2 = (\mathbf{r}, \mathbf{r})$, we have

$$rr' = (\mathbf{r}, \mathbf{v}) \quad \text{and} \quad rr'' + r'^2 = (\mathbf{r}, \mathbf{r}'') + v^2. \tag{4.19}$$

Thus, by using the angular momentum with the fundamental equality

$$r^2 v^2 = (\mathbf{r} \times \mathbf{v})^2 + (\mathbf{r}, \mathbf{v})^2 = c^2 + r^2 r'^2,$$

we have that

$$r'' = (\frac{\mathbf{r}}{r}, \mathbf{r}'') + \frac{v^2 - r'^2}{r} = (\frac{\mathbf{r}}{r}, \mathbf{r}'') + \frac{c^2}{r^3}. \tag{4.20}$$

For the inverse-p force law, Eq. 4.20 becomes

$$r'' = (\frac{\mathbf{r}}{r}, \mathbf{r}'') + \frac{c^2}{r^3} = (\frac{\mathbf{r}}{r}, \frac{-\mathbf{r}}{r^{p+1}}) + \frac{c^2}{r^3} = -\frac{1}{r^p} + \frac{c^2}{r^3}. \tag{4.21}$$

Observe from Eq. 4.21 how the angular momentum imposes different effects on the behavior of the system depending on the value of p. For instance, for $p < 3$ and for sufficiently small values of r, the c^2/r^3 term dominates the $-1/r^p$ term to serve as an effective repulsing force. It is this repulsing effect that denies the possibility of a collision and leads to the Weierstrass-Sundman theorem: this c^2/r^3 term plays a role similar to the repulsing term of the Lennard-Jones potential briefly discussed on page 44.

The story changes for $p \geq 3$. If $p > 3$, then the attracting gravitational term of $-1/r^p$ dominates the c^2/r^3 term for small r values, so, clearly, collisions can occur even with $c \neq 0$. The story is slightly more delicate for $p = 3$, where $r'' = (c^2 - 1)/r^3$, because of the need to impose the extra $c^2 < 1$ constraint: this condition means that collisions can occur only if the angular momentum is sufficiently weak. (The $c^2 = 1$ setting with $r'' \equiv 0$ is handled in the obvious manner.)

4.3.2 Larger p-values and "Black Holes"

As an amusing aside, recall those standard descriptions of "Black Holes" where anything that strays sufficiently close to the Black Hole gets sucked into a rapidly swirling descent. It is interesting how the above description shows that a similar swirling dynamic already occurs with far more elementary gravitational settings than relativity. In particular, a similar fate awaits any particle residing in a universe where the dynamics are governed by an inverse-p force law for $p \geq 3$.

First, notice from Eq. 4.21 that if a particle happens to be sufficiently close to the central force, then $r'' < 0$. This "sufficiently close" value is easy to compute: according to the right-hand side of Eq. 4.21, this radius is

$$r < \begin{cases} c^{-2/(p-3)} & \text{for } p > 3 \\ \infty & \text{for } c^2 < 1,\ p = 3. \end{cases} \tag{4.22}$$

According to this computation, if a particle ever strays into this region with $r' \leq 0$—this happens as r' must be negative for the particle to meander into this circular region—then the particle is doomed to perish in a collision. Stated in another manner, an open set of initial conditions have solutions that must terminate in collisions. But while the $\mathbf{r} \times \mathbf{v} = \mathbf{c} \neq \mathbf{0}$ integral is too weak to prevent collisions for $p \geq 3$ force laws, this angular momentum term remains in effect. It is easy to appreciate (and prove) what must happen: just as a child swinging a weight on a string discovers by pulling in the string, the combination of a particle satisfying this angular momentum relationship with $r \to 0$ ensures an interesting and rapidly increasing "spinning" effect on the collision behavior. Namely, the colliding particle experiences a spin growing so rapidly that it approaches infinite proportions.

This spinning effect extends beyond force laws that are strictly of the inverse $p > 3$ form to include classical Newtonian systems where the forces have a $-\frac{\mu}{r^2} + \frac{1}{r^p}$ form with $p > 3$. Just as with Eq. 4.20, once the r value becomes sufficiently small, it is the r^{-p} term of the force that dominates. These force laws are surprisingly common. After all, to handle perturbations caused by objects not being perfect homogenous spheres, a simplified version has potentials of the

$$\frac{\mu}{r} + \sum_{p=3} \frac{J_p}{r^p}$$

form where the J_p's are constants. (See, for example, almost any paper discussing artificial earth satellites.) This expression means that for values of r small enough, the larger values of p dominate forcing a twirling collision. However, the J values are sufficiently small and the radii of the planets are sufficiently large that such a death spin would not be observed. Instead, to create such an action, we would need a non-homogeneous planet (and, all of them are) with sufficiently large mass that has been shrunk to a small size—a description nicely agreeing with the black hole stories.

To complete this aside by returning to relativity, on page 43 it was pointed out that the relativity theory explanation for the advance of perihelion of Mercury can be described in terms of a $\frac{-1}{r^2} + \frac{\epsilon}{r^4}$ force law. Applying the above story to this force law leads to the same conclusion that if a particle ever strays sufficiently close to the central black hole source, then it will be sucked into a rapidly increasing, swirling descent. In other words, by combining the two stories about how force laws of the $\frac{-1}{r^2} + \frac{\epsilon}{r^4}$ form can arise, we learn that aspects of the black hole behavior—and relativity—already are captured by these classical potentials.

4.3.3 Lagrange-Jacobi equation

So far the argument suggests that complete collapse need not occur for force laws where $1 < p < 3$ should $\mathbf{c} \neq \mathbf{0}$. On the other hand, at least for the central force problem, the results of the preceding section prove for $p \geq 3$ that there are open sets of initial conditions where their solutions terminate in collisions. The next step is to explain for N-body settings why we must must expect complete collapse for $p \geq 3$ even if $\mathbf{c} \neq \mathbf{0}$.

An easy way to achieve this goal is to use an earlier introduced expression (Eq. 2.26 and proved on page 48) that is basic to our understanding about the mathematical behavior of collisions and singularities of the N-body problem: the Lagrange-Jacobi equation

$$I'' = U + 2h = 2T - U = T + h \tag{4.23}$$

where $\frac{1}{2}\mathbf{R}^2 = I = \frac{1}{2}\sum_{j=1}^{N} m_j r_j^2$ (page 40).

An interesting feature described in Cor. 2.1 (page 42) is that $\sqrt{2I} = R$ serves as a measure of the diameter of the N-body system while U^{-1} is a measure of the value of r_{min}. In other words, the $I'' = U + 2h$ form of the Lagrange-Jacobi equation introduces an interesting tension where a smaller value of r_{min} forces larger U and I'' values. Expressed in words, "close approaches" among particles accelerates the growth of the diameter of the system. Much of what we know about collisions and singularities relies on this interesting tension. But, what does this equation mean for general force laws? Some insight follows.

Generalized Lagrange-Jacobi

The Lagrange-Jacobi equation for the inverse-p force law, $p \neq 0$, is

$$I'' = 2T + (1-p)U = (3-p)U + 2h = (3-p)T + (p-1)h. \qquad (4.24)$$

Notice how this expression captures for N-bodies what we saw above for central force problems: vivid differences arise for the different force laws, particularly once $p \geq 3$. This is because if $p = 3$, then $I'' = 2h$, and we have the two extra integrals

$$I'(t) = 2ht + I'(0), \quad I(t) = ht^2 + I'(0)t + I(0).$$

Thus, if $h < 0$ and $p = 3$, we have the advertised assertion that $I \to 0$ in finite time; i.e., the system (if it lasts long enough) *must* suffer a complete collapse. For $p > 3$ and independent of the value of h, if some pair of particles closely approach each other (so U has a large value), expect collapse. (Compare this with comments in the above "black hole" section.) In other words, the importance of the Lagrange-Jacobi equation is that it replaces the r'' expression for the central force problem with an implicit equation for R'' for the N-body problem.[10]

The derivation of Eq. 4.24 is essentially that of Eq. 2.26, 4.23 or that in the footnote below. After differentiating I twice, we have

$$I'' = \sum_{j=1}^{N} m_j \mathbf{v}_j^2 + \sum_{j=1}^{N} \mathbf{r}_j \cdot m_j \mathbf{r}_j'' = 2T + \sum_{j=1}^{N} \mathbf{r}_j \cdot \frac{\partial U}{\partial \mathbf{r}_j}.$$

As U is homogeneous of degree $1 - p$, the last summation is $(1-p)U$, or

$$I'' = 2T + (1-p)U$$

where the rest of Eq. 4.24 follows from the energy integral.

4.3.4   Proof of the Weierstrass-Sundman Theorem

The plan of attack is to explain the Weierstrass-Sundman Theorem first for the central force problem, and then for the general N-body problem.

---

[10]To see this, differentiate $R^2 = \mathbf{R}^2$ twice to obtain $R''R + (R')^2 = <\mathbf{R}, \nabla_s U> +\mathbf{V}^2 = -U + 2T$. The middle and right-hand side of this expression proves Eq. 4.23 while the left-hand side is an expression involving R''. Thus we can and should treat the Lagrange-Jacobi equation as a disguised form of a R'' expression.

Central force problem.

While the Sect. 4.3.1 argument using Eq. 4.21 suggests why the Weierstrass-Sundman theorem should hold—at least for a central force problem—it is not a proof. The proof provided next for the central force problem, which emphasizes the energy integral, is surprisingly simple.

Theorem 4.2 *For the inverse p central force problem, $1 < p < 3$, collisions cannot occur if $c \neq 0$. Instead, r is bounded below by a constant that depends on the values of c and h. In particular, if $h < 0$ and $1 < p < 3$, then r is bounded both above and below by constants defined in terms of h and c.*

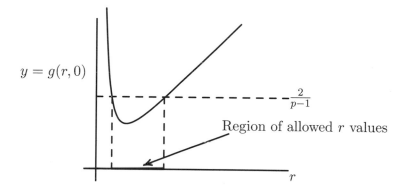

Fig. 4.2 Graph of $y = g(r, 0)$

Proof: By use of Eq. 4.19, the energy integral $\frac{1}{2}v^2 = (\frac{1}{p-1})\frac{1}{r^{p-1}} + h$ becomes

$$\frac{c^2}{r^2} + r'^2 = 2[(\frac{1}{p-1})\frac{1}{r^{p-1}} + h]. \tag{4.25}$$

Multiplying both sides of the equation by r^{p-1} and collecting terms leads to

$$g(r, r') = \frac{c^2}{r^{3-p}} + r^{p-1}r'^2 - 2hr^{p-1} = \frac{2}{p-1}, \tag{4.26}$$

or

$$g(r, 0) = \frac{c^2}{r^{3-p}} - 2hr^{p-1} \leq \frac{2}{p-1}. \tag{4.27}$$

For $1 < p < 3$ and $h < 0$, the graph of the left hand side of Eq. 4.27, $g(r, 0)$, is in Fig. 4.2. The portion of the graph on the left, with the y-axis as an asymptote, is dominated by the $\frac{c^2}{r^{3-p}}$ part of $g(r, 0)$, while the increasing portion of the graph to the right is governed by $|2h|r^{p-1}$. The horizontal dashed line represents the value $\frac{2}{p-1}$, so the values of r, as restricted by $g(r,0) \leq \frac{2}{p-1}$, are restricted to lie between the two vertical dashed lines. The actual upper and lower bound for r can be computed in a straightforward manner. This completes the proof for $h < 0$.

The only difference in the argument for $h \geq 0$ is that the right-hand tail of the $y = g(r,0)$ curve in Fig. 4.2 either asymptotically approaches zero (for $h = 0$), or it approaches negative infinity (for $h > 0$). Nevertheless, the left side of the curve asymptotically approaches infinity along the y-axis as $r \to 0$. This change in the graph merely means that this $g(r,0) \leq \frac{2}{p-1}$ constraint imposes lower bounds on r but no upper bounds—this is all that is asserted in the theorem. $\square$

Notice how different values of p make a difference. For instance, if $p > 3$, then Eq. 4.27 becomes $g(r,0) = c^2 r^{p-3} - 2hr^{p-1}$ where the graph approaches zero, rather than infinity, as $r \to 0$.

General N-body problem

I now prove the Weierstrass-Sundman Theorem for the general N-body problem rather than just for the three-body problem. Even with the extension from three to N particles given here, this proof is simpler than that of Weierstrass or Sundman and it suggests why $\sqrt{I}$ and R should be bounded away from zero if $\mathbf{c} \neq \mathbf{0}$ and $1 < p < 3$.

Theorem 4.3 *For $1 < p < 3$ and the general N body problem, it is impossible for $r_{max} \to 0$ if $\mathbf{c} \neq \mathbf{0}$.*

The approach is to extend the above central force argument to the general Newtonian N-body problem by replacing the r term with $\sqrt{I}$, or with R where $\mathbf{R}^2 = 2I$. The first goal is to generate an inequality of the Eq. 4.26 form. This proof uses the weaker form of the energy integral given by the Sundman's inequality Eq. 2.60. (As shown in the proof of Thm. 2.8 starting on page 61, Sundman's inequality is a special case of the velocity decomposition and energy integral where $\mathbf{W}^2_{config}$ is dropped.)

Proof: With the Sundman Inequality Eq. 2.60,

$$c^2 + (I')^2 \leq 4I(U + h), \tag{4.28}$$

solve for the *configurational measure* $I^{(p-3)/2}U$; i.e., subtract $4Ih$ from both sides, and then divide by $(\sqrt{I})^{p-3}$. In this manner, the N-body analogue of

Eq. 4.26 becomes

$$G(\sqrt{I}, I') = \frac{c^2 + I'^2}{(\sqrt{I})^{3-p}} - 4h(\sqrt{I})^{p-1} \le 4I^{(p-1)/2}U, \tag{4.29}$$

and Eq. 4.27 is replaced by the inequality

$$G(\sqrt{I}, 0) = \frac{c^2}{(\sqrt{I})^{3-p}} - 4h(\sqrt{I})^{p-1} \le 4I^{(p-1)/2}U. \tag{4.30}$$

This expression is the natural extension of Eq. 2.61, page 62 .

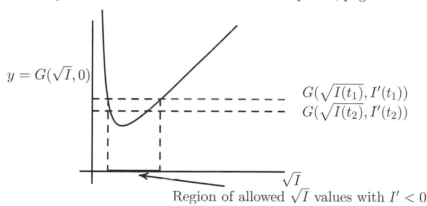

Fig. 4.3. Graph of $y = G(\sqrt{I}, 0)$

The Fig. 4.3 graph of $G(\sqrt{I}, 0)$ for $h < 0$ is essentially the same as the Fig. 4.2 graph of $g(r, 0)$: the minimum for $G(\sqrt{I}, 0)$ is at

$$\sqrt{I} - \sqrt{\frac{(3-p)c^2}{4(p-1)|h|}}. \tag{4.31}$$

Earlier (Thm. 2.2, page 50) we encountered this same $I = \frac{c^2}{4|h|}$ value for $p = 2$ in a different context. The reason this value keeps reappearing is suggested below with our discussion of G' where we show that the I value must be both above and below this $\frac{c^2}{4|h|}$ value.

Because $G(\sqrt{I}, 0)$ is bounded above by $4I^{(p-1)/2}U$ rather than a constant, the same argument proving the impossibility of $I \to 0$, or a complete collapse, does not hold. Instead, the value of this N-body configurational measure (which is homogeneous of degree zero) is determined by the shape of the configuration—but not its scale or orientation.[11] To sidestep the re-

---

[11]As the *configurational measure* $I^{(p-1)/2}U$ is key for finding central configurations (Eq. 2.20, page 41), it adds support for earlier assertions (Sect. 2.2.1, page 36) that central configurations play a role in the study of collisions. This is the case.

ality that the configurational measure is not a constant for $N \geq 3$, we need a replacement argument: the first step is the following Lemma.

Lemma 4.1 *For $p \neq 0$, the expression*

$$G'(\sqrt{I}, I') = (3 - p)\frac{I'}{2I^{(5-p)/2}}[-(c^2 + I'^2) + 4IT] \tag{4.32}$$

holds. In particular, for $p < 3$, if $I' > 0$, then $G' \geq 0$; if $I' < 0$, then $G' \leq 0$. For $p > 3$, these inequalities are reversed; i.e., if $I' > 0$, then $G' \leq 0$.

A slightly stronger assertion follows from Thm. 2.10: as the term in the brackets in Eq. 4.32 is zero iff the particles are in the invariable plane and $\mathbf{W}_{config} = \mathbf{0}$, we can be precise about when $G' = 0$ but $I' \neq 0$.

Proof of Lemma 4.1. Differentiation shows that

$$\begin{aligned}
G'(\sqrt{I}, I') &= \frac{(p-3)(c^2+I'^2)}{2I^{(3-p)/2}} + \frac{2I'I''}{I^{(3-p)/2}} - 2h(p-1)\frac{I'}{I^{3-p)/2}} \\
&= (3-p)\frac{I'}{2I^{(5-p)/2}}[-(c^2 + I'^2) + 4I(I'' - (p-1)h)].
\end{aligned}$$

But according to the generalized Lagrange-Jacobi equation Eq. 4.24, we have $4I(I'' - (p-1)h) = 4IT$; this verifies Eq. 4.32.

The rest of the lemma, which requires the growth of G to be coordinated with that of I, follows from the Sundman inequality as given in Eq. 2.60, 4.28. Namely, the term in the brackets of Eq. 4.32 is non-negative, so the signs of I' and G' cannot differ if $p < 3$, and they cannot agree if $p > 3$. □

To complete the proof of Thm. 4.3, replace the $2/(p-1)$ constant in Eq. 4.27 with the growth behavior of $G'(\sqrt{I}, I')$. In particular, if it were possible for $I \to 0$ as $t \to t^*$, then at some time $t_1 < t^*$, we have $I'(t_1) < 0$. In turn, Lemma 4.1 tells us that as long as $I' \leq 0$, we have that

$$G(\sqrt{I}, 0) \leq G(\sqrt{I(t_1)}, I'(t_1)).$$

Thus, during this time period when $I' \leq 0$, the value of I is strictly bounded below. To find this lower bound for I, as depicted in Fig. 4.3, the value $G(\sqrt{I}, I')$ must lie below the higher horizontal dashed line. Thus the lower bound for I is where this line meets the left-hand branch of G.

Consequently, if it were possible for $I \to 0$ as $t \to t^*$, it could not occur monotonically with $I' < 0$. Instead I' must be positive for some time interval. Suppose this occurs where I' changes sign at t_2, $t_1 < t_2 < t^*$. But as $I'(t_2) = 0$, $G(\sqrt{I(t_2)}, I'(t_2))$ must be on the curve $y = G(\sqrt{I}, 0)$, and I must increase for a time interval after t_2. Again according to the lemma, for all

$t > t_2$ where $I'(t) > 0$, it must be that $G(\sqrt{I(t)}, I'(t)) \geq G(\sqrt{I(t_2)}, I'(t_2))$: the graph of G must be above the bottom dashed line of Fig. 4.3. This last inequality, along with the fact that whenever $I' = 0$ the value of $G(\sqrt{I}, I')$ must be on the graph of $G(\sqrt{I}, 0)$, means that there is a lower bound on the value of I before I can decrease again. In particular, the next time $I' = 0$, the value of $\sqrt{I}$ must be at least equal to where the lower dashed line in Fig. 4.3 meets the right-hand branch of G.

As this argument holds for any $I(t_2)$ and as the next critical point of I must be where $I > \frac{(3-p)c^2}{4(p-1)|h|}$ (Eq. 4.31), it follows that it is impossible for $I \to 0$ if $\mathbf{c} \neq \mathbf{0}$. This completes the proof. $\square$

The last part of this proof evokes the image where a singularity with $\mathbf{c} \neq \mathbf{0}$ and $h < 0$ might replace a complete collapse with the wild action where I increases to a large enough value to allow a small subsequent I value. According to the above argument and Fig. 4.3, this motion would require

$$\liminf_{t \to t^*} I = 0, \quad \limsup_{t \to t^*} I = \infty. \tag{4.33}$$

Such behavior would be delightfully perverse, but it follows from Sect. 4.4 that the motion cannot happen. Instead, as established in Marchal and Saari [40] for the standard Newtonian inverse-square force law, with $\mathbf{c} \neq \mathbf{0}$, once I sinks below a particular value, it must be that $I \to \infty$. The idea of the proof is that a sufficiently small value of I requires the dashed line to be arbitrarily high in Fig. 4.3, and $G(\sqrt{I}, I')$ must be above this line. By taking more care with the properties of $G(\sqrt{I}, I')$, it follows that it can never hit the right branch of $G(\sqrt{I}, 0)$, so the motion expands to infinity. In this way, Sundman's result for the three body problem, where he established a lower bound on how close I can approach zero and a result by Birkhoff for the three-body problem showing that a sufficiently small value of I requires $I \to \infty$, are extended to the general N-body problem.

4.3.5   Bounded above means bounded below

Before moving on, let me digress to extract an immediate consequence of the above argument: this result captures the flavor of the preceding paragraph describing how sufficiently small R values force the subsequent N-body behavior to have $R \to \infty$. To motive the kind of results that can be obtained, notice how the argument for the central force law allow a stronger conclusion: the Fig. 4.2 graph of $y = g(r, 0)$ and the $2/(p-1)$ value impose upper and lower bounds on the value of r for $h < 0$. While this strong conclusion fails to hold for the N-body problem, some of the flavor does. Namely, if the diameter of the universe, R, is bounded above for all time, then R is bounded strictly away from zero.

Our proof uses a fact that probably has been known for centuries (at least for $p = 2$) and that has been implicit in several earlier comments.

Proposition 4.1 *If a solution for the N-body problem, where $1 < p < 3$, exists for all time and $h \geq 0$, then r_{max} and R are unbounded. Namely, bounded motion can occur only if $h < 0$.*

Proof: This assertion is an immediate consequence of the the Lagrange-Jacobi relationship Eq. 4.24, which is $I'' = (3 - p)U + 2h$. If $h > 0$, then $I'' > 2h$, so, if the solution lasts long enough, $I > ht^2 + I'(0)t + I(0)$, which means that the motion is unbounded. For $h = 0$, use the fact, which follows from $r_{min}^{-1} \geq r_{max}^{-1}$ and Cor. 2.1 (page 42), that there is a positive constant A depending on the masses so that[12]

$$U \geq \frac{A}{I^{(p-1)/2}}.$$

So, when $h = 0$, the Lagrange-Jacobi equation becomes

$$I'' \geq \frac{A}{I^{(p-2)/2}}.$$

Now, if I were bounded for all time, it follows from the last inequality that I'' would be bounded away from zero for all time, or that $I \to \infty$ as $t \to \infty$. This contradiction proves that I must be unbounded when $h = 0$. . $\square$

This assertion is used to prove that if R is bounded above, it must be bounded below.

Theorem 4.4 *For the N-body problem and $1 < p < 3$, if $\mathbf{c} \neq \mathbf{0}$ and if there is a value D so that $I(t) \leq D$ for all time, then there is a value $D^* > 0$ so that $I(t) \geq D^*$ for all time. Indeed, D^* satisfies*

$$G(\sqrt{D^*}, 0) = G(\sqrt{D}, 0).$$

If $D > \frac{c^2}{4|h|}$, then $D^* < \frac{c^2}{4|h|}$. If $D = \frac{c^2}{4|h|}$, then $D^* = \frac{c^2}{4|h|}$ and the motion is a relative equilibria solution.

---

[12]This inequality states that the configurational measure $R^{p-1}U$ must be bounded below. This assertion follows immediately from the weighted mean arguments of Chap. 2; e.g., one lower bound is found by replacing all $r_{j,k}$ values with unity.

Proof: As asserted in Prop. 4.1, bounded motion requires $h < 0$. If there is a value of D so that $I(t) \leq D$ for all time, this D value defines a horizontal line $y = G(\sqrt{D}, 0)$ on Fig. 4.3 that passes through $y = G(x, 0)$ in two values: one is $x = \sqrt{D}$ and the other defines $x = \sqrt{D^*}$.

The conclusion is immediate for a constant $R = 2I$. If I is not a constant, then the argument developed above shows that when I is decreasing in value (from $I(t_0) \leq D$ for some t_0), $G(\sqrt{I}, I')$ must be below the horizontal line $y = G(\sqrt{D}, 0)$, so $I(t) \geq D^*$.

The bound on D^* and D come from Fig. 4.3 and the fact that $\sqrt{\frac{c^2}{4|h|}}$ is the minimum point for G. The assertion that if $D = \frac{c^2}{4|h|}$, then the motion is a relative equilibria comes from the fact that $I \equiv \frac{c^2}{4|h|}$ and Thm. 2.2. This completes the proof. $\square$

4.3.6 Problems

Much remains to be learned. For instance, while considerable amount of the study of collisions and singularities has concentrated on their properties, one must wonder how the location of the complex singularities are related to the behavior of the N-body problem. It is reasonable to question, for instance, whether the location of complex singularities are, in an manner, related to the minima of the minimum spacing between particles. If all particles expand to infinity, for instance, will the radius of convergence also become arbitrarily large? What is the relationship between the radius of convergence and certain behaviors of the Newtonian N-body problem? More generally, what role do locations and properties of the complex singularities play in the general behavior of the N-body problem? Interesting questions: while I derived some results many years ago, I know of no systematic investigations in these directions.

In Scct. 4.4, "noncollision singularities" are introduced and discussed. As described then, the minimum spacing between particles goes to infinity much faster than the $(t^* - t)^{2/3}$ value, and it must be accompanied by a highly oscillatory behavior of the system. While this motion is *not* the limit of real binary collisions, I conjecture that it is the limit of complex singularities. To the best of my knowledge, nobody has explored this question.

4.3.7 An interesting historical footnote

Before providing an overview of singularities and a discussion of the properties of collisions, let me include a historical aside describing how celestial mechanics has played an important role in the development of some classical results from complex variables. As mentioned above, even the elliptic solution for the simple central force problem does not have a closed form representation. Instead, this solution is defined implicitly through Kepler's equations Eq. 4.17.

To avoid the problems of having only an implicit solution, during various periods of history mathematicians sought a useful direct expression; this included even trying to find a power series solution for r. As Wintner [112] asserts in a footnote on page 217 of his book,

> *"The direct proof of [this aspect of Kepler's equation] played an important historical role in the theory of analytic functions ...[A] principal impetus for Cauchy's discoveries in complex function theory was his desire to find a satisfactory treatment for Lagrange's series [for Kepler's equation]. Cauchy was led to his fundamental theorem connecting the radius of convergence with the location of the nearest singularity, as well as to his maximum principle, precisely in his papers dealing [with this problem]. Also the facts usually referred to as the argument principle and Rouche's theorem were first observed in connection with this problem concerning Kepler's equation."*

4.4   Singularities – an overview

What is a singularity for the Newtonian N-body problem? Can the deliciously wild behavior suggested by Eq. 4.33, where the radius of the system expands and contracts in a radical manner, occur? To address this and other questions, the notion of a singularity for the N-body problem needs to be introduced. Everything follows from the existence theorems of differential equations that describe when solutions exist.

To review, recall that with a given differential equation and an initial condition,

$$\mathbf{x}' = f(\mathbf{x}), \quad \mathbf{x}(t^*) = \mathbf{x}_0,$$

standard existence theorems assert that if f is smooth and $||f(\mathbf{x})|| \leq M$ in a neighborhood $||\mathbf{x} - \mathbf{x}_0|| < a$ of the initial condition $\mathbf{x}_0$, then a unique solution exists for the system at least for $|t - t^*| < b$. The important point is that the value of b is strictly determined by the values of a and M.

To extend a solution beyond this local conclusion, it is standard to use analytic continuation. Namely, if a solution exists for $|t - t^*| < b$, then for a value of t near (or even on) the boundary of this region, reapply the existence theorem. Continue this approach as long as possible to define the maximal region of definition. A *singularity* of the system is a time value $t = t^*$ where analytic continuation fails; i.e., the solution cannot be continued beyond t^*.

It is clear from the equations of motion for the N-body problem, $\mathbf{r}''_j = \frac{\partial U}{\partial \mathbf{r}_j}$ (see Eq. 2.5), that to have a singularity at $t = t^*$, some mutual distance $r_{j,k}$ must become arbitrarily small as $t \to t^*$. Trivially, a collision, which is defined next, is a singularity.

Definition 4.1 *A collision at time $t = t^*$ occurs if each $\mathbf{r}_j$ approaches a distinct limit as $t \to t^*$ and at least two of these limits agree.*

4.4.1 Behavior of a singularity

Collisions are singularities, but can a singularity be cause by other kinds of behavior? Could, for instance, two particles try to collide but in doing so they spin arbitrarily fast around some circle and never meet? A wilder scenario is where the system flirts with the sense of a collision but, at the last instant, the particles change their minds so they never hit. To be more specific, can the limit inferior of

$$r_{min}(t) = \min_{j \neq k} r_{j,k}(t)$$

approach zero while the limit superior remains bounded away from zero?

These kinds of existence questions bothered the area of celestial mechanics during the nineteenth century. During his 1895 Swedish lectures [56], P. Painlevé[13] characterized the behavior of singularities.

Theorem 4.5 *(Painlevé [56]) A singularity for the N-body occurs at time $t = t^*$ if and only if $r_{min}(t) \to 0$ as $t \to t^*$.*

---

[13]Paul Painlevé (1863-1933) received the significant recognition and national honor of being interned in the Pantheon in Paris. Not for his mathematics, but for his important contributions to French *politics* including *twice* serving as the French Prime Minister. As for his mathematical career, while he was interested in rational transformations of algebraic curves and surfaces, he probably is better known for his work on mechanics and, in particular, his study of differential equations and their singularity structure along with the Painlevé transcendents. Among his mathematical awards is the Grand Prix des Sciences Mathématique (1890), the Prix Poncelet (1896), and his election to the Académie de Sciences. An unusual tidbit related to his interest in aviation is that in October 28, 1908, he was Wilbur Wright's first French passenger for a 70 minute flight! As increasingly true with flights today, I doubt that snacks were served. (Wright's first passenger, only five months before in Kitty Hawk, was C. Furnas. The month before Painlevé's flight, Orville Wright's plane crashed and killed the passenger Lt. T. Selfridge.)

Proof: The proof is a direct application of the existence theorem for differential equations. Suppose the conclusion is false because there is a singularity at $t = t^*$ where

$$\limsup_{t \to t^*} r_{min}(t) = D > 0.$$

If so, there is a sequence $\{t_j\}$, $t_j \to t^*$, where $r_{min}(t_j) \to D$. Eventually (i.e., for sufficiently large choices of j) we have that $r_{min}(t_j) > D/2$ at each t_j, so it follows from the inverse-square form of the equations that for each k all of the $\|\frac{\partial U}{\partial \mathbf{r}_k}\|$ and U are bounded above in a neighborhood about $\mathbf{r}(t_j) = (\mathbf{r}_1(t_j), \ldots, \mathbf{r}_N(t_j))$ where $r_{min} > a = D/4$. Moreover, this bound is in terms of the value of D and the masses.

According to the conservation of energy integral (Eq. 2.6, page 34), if U is bounded, then so are each of the $\mathbf{v}_k$ variables. Consequently in a neighborhood about the solution for each t_j, the system

$$\mathbf{r}'_k = \mathbf{v}_k, \quad \mathbf{v}_k = \frac{\partial U}{\partial \mathbf{r}_k}$$

is bounded where the bound is in terms of D, the masses, and h.

According to the existence theorem, there is a value $b > 0$, which depends only on D and h, so that the solution exists in a $|t - t_j| < b$ neighborhood. But $t_j \to t^*$, so eventually we have that $|t_j - t^*| < b/2$. This inequality implies that t^* is a regular point, not a singularity. This contradiction proves the assertion. $\square$

To describe Painlevé's conclusion in a geometric manner, let $\Delta_{j,k}$ (where "Δ" intended to suggest "diagonal") be defined as

$$\Delta_{j,k} = \{\mathbf{R} = (\mathbf{r}_1, \ldots, \mathbf{r}_N) \in (\mathbb{R}^3)^N \mid \mathbf{r}_j = \mathbf{r}_k\}$$

and

$$\Delta = \cup_{j \neq k} \Delta_{j,k}.$$

Painlevé's result means that the system suffers a singularity at $t = t^*$ iff $\mathbf{R}(t) \to \Delta$ as $t \to t^*$. This requirement introduces two classes of singularities. The first is a *collision*, which is defined (page 163) to be where $\mathbf{R}$ approaches a *specific point* in Δ. But Painlevé's condition allows for the possibility that $\mathbf{R}$ approaches the set Δ without approaching any specific point in Δ. In other words, as Painlevé recognized, his condition allows for the possibility of a *non-collision singularity*—a singularity of the system that is not a collision. At least for now, this claim includes the possibility of one body chasing another in a rapid spin along a circle among other wild choices.

4.4.2 Non-collision singularities

Do non-collision singularities exist? Painlevé proved that they do not for the three-body problem. His argument uses the Lagrange-Jacobi equation $I'' = U + 2h$ of Eq. 4.23. I leave it as a simple exercise for the reader to use the extended Lagrange-Jacobi equation Eq. 4.24 to extend the following proof to other inverse-p force laws $1 < p < 3$.

Theorem 4.6 *(Painlevé [56]) For the three-body problem, all singularities are collisions.*

Proof: If there is a singularity at $t = t^*$, then $r_{min}(t) \to 0$ as $t \to t^*$. In turn (Eq. 2.23, page 42), when $r_{min}(t) \to 0$, then $U \to \infty$ as $t \to t^*$ and, according to the Lagrange-Jacobi relationship Eq. 4.23, $I'' \to \infty$. Thus, as $I'' > 0$ for t sufficiently close to t^*, I must approach a limit; e.g.,

$$I \to D \text{ as } t \to t^*.$$

If $D = 0$, then all particles must collide at the center of mass and we are done; i.e., each $\mathbf{r}_j \to \mathbf{0}$. So assume that $D > 0$. (At this stage there is nothing to preclude the possibility that $D = \infty$.) The first goal is to establish that, eventually, the same pair of particles defines r_{min}.

As $t \to t^*$, the "$I \to D > 0$" requirement ensures that the maximum spacing between particles eventually remains larger than some specified positive number, say E. (See Eq. 2.22, page 42.) But $r_{min} \to 0$, so eventually $r_{min}(t) < E/2$. Thus, eventually at each subsequent instant of time, at least one of the leg lengths satisfies $r_{j,k}(t) < E/2$.

It now follows from the triangle inequality that eventually the *same pair* of particles defines r_{min}. If this were not the case, then when two pairs, say $r_{1,2}$ and $r_{2,3}$, trade the role of being the minimum, we would have $r_{1,2} = r_{2,3} < E/2$, which would contradict the triangle inequality contradiction because $r_{1,3} > E$.

It remains to prove that all particles approach definite limits as $t \to t^*$. One particle, say particle 3, eventually is bounded away from the other two, so the equations of motion ensure there is a constant A so that

$$|\mathbf{r}_3''| < A \text{ as } t \to t^*.$$

Integrating from t_1 to t_2 establishes that

$$|\mathbf{v}_3(t_2) - \mathbf{v}_3(t_1)| < A(t_2 - t_1) \to 0 \text{ as } t_1, t_2 \to t^*. \tag{4.34}$$

Thus, according to the Cauchy criterion for the existence of a limit, $\mathbf{v}_3$ has a finite limit as $t \to t^*$.

It now follows that $\mathbf{r}_3$ also approaches a definite limit $\mathbf{L}_3$. The center of mass integrals (Eq. 2.2, page 33) and the fact that $r_{1,2} \to 0$ shows that $\mathbf{r}_1, \mathbf{r}_2 \to -\frac{m_3}{m_1+m_2}\mathbf{L}_3$ as $t \to t^*$. This completes the proof. $\square$

Notice how the above proof establishes that I must be bounded at the collision. This comment follows from Eq. 4.34, the finite value of A and the fact that the collision occurs in finite time. The reason for noting that I is bounded will become clear in the next section.

4.4.3   Off to infinity

The general situation involving $N > 3$ bodies is more complicated. What seemed to close the book on this topic was the following stunning result found in 1908 by von Zeipel.[14]

Theorem 4.7 (*von Zeipel [110]*) *A singularity at $t = t^*$ is a collision iff I is bounded as $t \to t^*$.*

Of course, as $r_{min}(t) \to 0$ and $I'' \to \infty$ as t approaches a singularity, we have that I must approach a limit for singularities. So, if I is not bounded, then $I \to \infty$ as $t \to t^*$. In other words, a non-collision singularity occurs iff the N-body system expands to infinity in finite time! How can that happen? This bizarre requirement, which is seemingly impossible, probably caused interest in this question to remain dormant for a half century. It was resurrected in the mid-1960s as part of the characterization of the behavior of collisions (Saari [72]); this topic is described starting in Sect. 4.5.

Close approaches

In order for $I \to \infty$ in finite time, surprisingly close approaches of particles are required. To explain why this is so, in the discussion of collisions given a bit later, one of the main results is that colliding particles must approach one another like $(t^* - t)^{2/3}$. In turn, this means that $U \sim A(t^* - t)^{-2/3}$ for some

---

[14]Edvard Hugo von Zeipel (1873-1959) was a professor of astronomy at Uppsala University in Sweden where he worked on theoretical issues in celestial mechanics and astrophysics. While those of us in celestial mechanics know that the "von Zeipel theorem" is Thm. 4.7, our friends in astrophysics consider the "von Zeipel theorem" as involving a state of radiative equilibrium in spheriods. Indeed, this aspect of von Zeipel's work arises often in studies of the evolution of rotating stars, etc. It is not surprising that, in 1992, an asteroid was named in his honor.

positive constant A. Substituting this relationship into the Lagrange-Jacobi equation Eq. 4.23 provides an integrable relationship $I'' \sim A(t^* - t)^{-2/3}$; by being integrable, it ensures a bounded I.

Turning this integrability observation around, the Lagrange-Jacobi equation clearly requires that for I to become unbounded, U must approach infinity much more rapidly than allowed by collisions: U must grow fast enough to prevent having an integrable I''. In turn, this requires the particles to approach one another much more rapidly than allowed by collisions.

Indeed, a necessary and sufficient condition for a singularity to be a collision (a condition weaker than von Ziepel's but much easier to prove) is that I' is bounded as $t \to t^*$. That this condition is necessary follows from the above assertions proved later in this chapter describing the asymptotic behavior of $\mathbf{r}_j$ and $\mathbf{v}_j$ as a collision is approached: sufficiency is established next.

Theorem 4.8 *(Pollard and Saari [64]) A necessary and sufficient condition for a singularity at time $t = t^*$ to be a collision is that I' remains bounded as $t \to t^*$.*

Proof: By integrating the Lagrange-Jacobi equation Eq. 4.23 from t to $a < t^*$, we obtain

$$I'(a) - I'(t) = \int_t^a U(\tau)\, d\tau + 2h(a - t).$$

With the assumption that I' is bounded and a positive integrand for the integral, it follows by letting $a \to t^*$ that the integral $\int_t^{t^*} U(\tau)\, d\tau$ converges. According to the Schwarz inequality, the integral $\int_t^{t^*} \sqrt{U(\tau)}\, d\tau$ must also converge.

A singularity requires $r_{min} \to 0$, which means that $U \to \infty$ as $t \to t^*$. So, eventually (i.e., as $t \to t^*$), the kinetic energy T satisfies $T \le 4U$. As each $\mathbf{v}_j$ has the property that $|\mathbf{v}_j| < A\sqrt{T}$, eventually we have that

$$|\mathbf{r}_j(t_2) - \mathbf{r}_j(t_1)| \le A \int_{t_1}^{t_2} \sqrt{T(\tau)}\, d\tau < 2A \int_{t_1}^{t_2} \sqrt{U(\tau)}\, d\tau \text{ as } t_1, t_2 \to t^*.$$

The integral on the right-hand side is convergent, so the Cauchy criterion ensures that each $\mathbf{r}_j(t)$ approaches a limit as $t \to t^*$. $\square$

Oscillations

A result stronger than von Zeipel's theorem is the following Thm. 4.9. Of particular interest, its proof shows that a non-collision singularity requires particles to approach each other in an oscillatory fashion infinitely often: while a singularity requires that $\mathbf{R} \to \Delta$, a non-collision singularity requires $\mathbf{R}$ to infinitely often approach, and then leave, certain $\Delta_{i,j}$ leafs.

Theorem 4.9 *(Saari [79]) If a singularity of the N-body problem at $t = 0$ has I is slowly varying as $t \to 0$, the singularity is due to collisions.*

To explain the terminology, Feller [20] (pp. 272-276) defines a function f to be "slowly varying" as $t \to 0$ if $f(\beta t)/f(t) \to 1$ for an arbitrary positive constant β. Clearly, if f approaches a finite non-zero limit as $t \to 0$, then f is slowly varying. An example of a slowly varying function that goes to infinity as $t \to 0$ is $(\ln t)^5$. What I show next is that von Zeipel's result, Thm. 4.7, is a special case of the above Thm. 4.9

Proof: To show that Thm. 4.7 is a corollary of Thm. 4.9, we need to show that if I is bounded as $t \to t^*$ (where t^* is a singularity), then I is a slowly varying function. The argument is immediate: as $t \to t^*$, $U \to \infty$ and, according to the Lagrange-Jacobi equation, $I'' \to \infty$. As I'' eventually is positive, we have that I is concave up. Consequently, I either approaches a non-negative limit (non-negative because $I \geq 0$), or I approaches infinity. The assumption that $I = O(1)$ rules out the $I \to \infty$ option. If $I \to 0$, then each $\mathbf{r}_j \to \mathbf{0}$, so the singularity is a collision. The remaining case is that $I \to L > 0$. But by approaching a finite limit, I is slowly varying. $\square$

To avoid disrupting the flow of the discussion, the proof of Thm. 4.9 is deferred to Sect. 4.9 (page 195). But to see why such motion must be accompanied by highly oscillatory motion, notice that if there an $\epsilon > 0$ can be found to bound particle $\mathbf{r}_i$ away from the others, then $|\mathbf{r}_i''| < A/\epsilon^2$. This bound may be very large, but there is only a finite amount of time until the singularity. Consequently, by integrating twice and using the Cauchy criterion, it follows that $\mathbf{r}_i$ approaches a limit.

Similarly, particles that are near each other form a natural grouping. To be a non-collision singularity, particles from one group have to intermingle, arbitrarily closely, with particles from another group. To see why this is so, compute the center of mass of a group: the acceleration of this center of mass is determined by the proximity of particles *outside* of the group. If the intermingling does not occur, then the acceleration of center of mass is bounded above, and, with the limited time until a singularity, the center of mass approaches a limit. In such a setting, it is not difficult to show that the singularity is a collision.

The proof of Thm. 4.9 partitions the particles into different groupings: each group has particles that are "close together," at least for a while, while the distance between groups is distant. Thus, the acceleration on the centers of mass of these groups is very small. But, as suggested, particles from one group must eventually intermingle with particles from another group to avoid the singularity being a collision. The dynamic of at least one particle moving from one group to another requires the associated center of mass to change—radically and quickly. The proof of the theorem shows that if the asymptotic behavior of I is that of a slowly varying function, the centers of mass cannot grow fast enough to permit the necessary dynamic.

Mather-McGehee

Such a scenario requiring rapid close approaches that are carried out in a manner so that particles go off to infinity in finite time seems to be impossible. Is it? Any sense of skepticism about the existence of such motion disappeared with the publication of a beautiful paper by John Mather and Richard McGehee [42] where they proved for the collinear four-body problem that binary collisions can accumulate[15] at a time t^*. Moreover, the motion is such that $I \to \infty$ as $t \to t^*$. The much faster growth rate required of U comes from the rapidly decreasing time intervals between successive collisions. This paper is a consequence of the "blow-up" construction McGehee developed [44, 45] to understand behavior near a triple collision: I strongly recommend these papers to the reader .

OK, so Mather-McGehee argument fails to be a non-collision singularity because it requires an infinite number of binary collisions. But their seminal result suggests that such motion is possible. Indeed, Anosov [3] suggested that if a non-collision singularity does exist, it probably could be found in a neighborhood of the Mather-McGehee orbit. This intriguing suggestion has not been verified.

Zhihong (Jeff) Xia

While this topic was hotly pursued, it seemed unlikely to be resolved at any time in the near future. *But, the question was finally answered: motion allowing particles to go off to infinity in finite time does exist.* In one of the more important contributions to the Newtonian N-body problem, one of my graduate students, Zhihong (Jeff) Xia, proved in his Ph. D. thesis [113] that such motion exists for the five-body problem!

---

[15]On page 161 I conjectured that a non-collision singularity is the accumulation of imaginary collisions that occur at imaginary values of time. This conjecture is motivated by the Mather-McGehee construction and the location of complex singularites relative to the motion of particles.

Fig. 4.4. Xia's construction

A complete and careful description of Xia's construction would require a separate book: for now his paper [114] is an excellent reference. To provide a sense of his proof, Xia has two pairs of particles in elliptic orbits about the z-axis and parallel to the x-y plane; one pair is above the plane and the other is below. A fifth particle commutes between the two pairs. Imagine the scenario where a commuting particle approaches a pair just prior to the pair reaching its closest approach. Then, once the commuting particle has just passed past the pair, the pair is at its closest approach that allows it to exert a much stronger pull on the commuter. This added acceleration flings the commuter back toward the other pair where the scenario is repeated. For a more complete expository description, see Saari and Xia [95].

Thanks to Xia's theorem, we now know that non-collision singularities can occur for $N \geq 5$, and we know from Painlevé that they cannot occur for $N \leq 3$. The remaining open question is to determine whether such motion can occur for $N = 4$. Maybe an answer can be found by carrying out Anosov's suggestion given above; maybe a direct proof is needed. But for readers interesting in tackling this question, be warned that while fascinating, the problem is very difficult!

4.4.4   Problems

This overview of N-body singularities leads to several other unresolved questions. Two are given here; the first concerns the structure of the sets; the second involves another approach used to construct non-collision singularities.

As described above, a non-collision singularity requires an infinite number of arbitrarily close approaches. The construction, then, appears to be very delicate because it is reasonable to expect that the slightest change in initial conditions could cause these closely approaching particles to actually collide. The natural question, then, is to find the relationship between collisions and non-collision singularities. Namely, if $\mathcal{C}$ consists of the set of initial conditions leading to some collision, and if $\mathcal{NC}$ is the set of initial conditions leading to a non-collision singularity, it is reasonable to question the structure of these two sets. In particular, is it true that

$$\mathcal{NC} \subset \overline{\mathcal{C}} \tag{4.35}$$

where $\overline{\mathcal{C}}$ is the closure of $\mathcal{C}$? Even stronger, is it true that

$$\mathcal{C} \cup \mathcal{NC} = \overline{\mathcal{C}}? \tag{4.36}$$

Only tentative results have been obtained by Saari and Xia [96], and I have doubts whether Eq. 4.36 is true. For instance, there may be orbits that flirt enough with collisions to be in $\overline{\mathcal{C}}$ but are not non-collision singularities.

A different question is based on an alternative attempt to construct a coplanar example of a non-collision singularity. With a clever use of symmetry, J. Gerver [21] created a $3N$-body problem where N pairs of particles, each with mass equal to unity, rotate about each other centered at the symmetry vertices of an expanding N-gon. The remaining N particles circulate, in a symmetric manner, from one pair to the next. The three-body interactions of each commuting particle with a pair has two close approaches—the particle whips about one mass only to just catch the other mass from a pair to whip about it. Much like running down the street and swinging around a gate to pick up speed, this "Z" shaped orbit adds speed to particle. Gerver asserts that with a large enough value of N and a small enough value for the common mass, m, of the commuting particles, the solution creates a non-collision singularity.

While I applaud Gerver's clever construction, his paper is difficult to read. A valued contribution would be to reconstruct Gerver's proof in a more geometric sense. By doing so, these unresolved values of N and m probably could be established. Even more; my sense is that by making his proof more transparent, an uncountable number of new behaviors might be found. Namely, the "Z" interactions probably could be replaced with a wide class of interactions captured by some form of symbolic dynamics. The first step in carrying out this program is to extend McGehee's [44] construction for the behavior of near triple collisions from the collinear three body problem to the coplanar three-body problem.

4.5   Rate of approach of collisions

I now discuss the behavior of collisions by first determining their rate of approach. Binary collisions were handled by Sundman [106, 107] where he established the existence of a constant $A_{j,k}$ for which

$$r_{j,k} \sim A_{j,k}(t^* - t)^{2/3}. \tag{4.37}$$

Triple collisions in the three-body problem were handled by C. L. Siegel [100] establishing the same rate of approach. Around the same time, A. Wintner [112] showed that Eq. 4.37 holds for total collapse in the N-body problem.

What remained was to analyze general collisions; e.g., maybe there is a five-body collision in the ten-body problem, or maybe the 50-body problem suffers simultaneous collisions where a triple collision occurs in one location, a five-body collision occurs elsewhere, a ten-body collision occurs in a third location, and the remaining 32 bodies are free from these collisions. The behavior of general collisions was resolved in my thesis [72]. After Pollard and I improved on my arguments and exposition, we [64] published Thm. 4.10 given below.

4.5.1   General collisions

For notation, as a collision occurring at time t^* requires each particle to approach a definite limit, let them be $\mathbf{L}_1, \ldots, \mathbf{L}_N$. Partition the indices of the particles according to their limit (i.e., according to who collides with whom) where

$$G_j = \{k, i \,|\, \mathbf{L}_i = \mathbf{L}_k\}.$$

To illustrate with the above scenario for $N = 50$, there is a five-body collision at $\mathbf{L}_1 = \cdots = \mathbf{L}_5$, a triple collision at $\mathbf{L}_6 = \mathbf{L}_7 = \mathbf{L}_8$, and a ten-body collision at $\mathbf{L}_9 = \cdots = \mathbf{L}_{18}$, so sets G_1, G_2, G_3 would consist of, respectively, three, five and ten indices, and the remaining 32 sets of G_j are singletons.

Replacing I, the polar moment of inertia, is the term

$$J = \frac{1}{2} \sum_{j=1}^{q} \sum_{k \in G_j} m_k (\mathbf{r}_k - \mathbf{L}_k)^2, \tag{4.38}$$

which captures the rate of approach of colliding particles. Notice how this expression separates each colliding system into its own "complete collapse." The non-colliding particles just approach their limit.

Theorem 4.10 *(Pollard and Saari [64]). A necessary and sufficient condition for a singularity at time $t = t^*$ to be a collision is if there is a positive constant A so that*

$$U \sim \frac{4}{9} A (t^* - t)^{-2/3} \text{ as } t \to t^*. \tag{4.39}$$

With a collision at time $t = t^*$, there is a positive constant A so that

$$\begin{aligned} J &\sim A(t^* - t)^{4/3} \\ J' &\sim -\tfrac{4}{3} A (t^* - t)^{1/3} \quad \text{as } t \to t^* \\ U &\sim \tfrac{4}{9} A (t^* - t)^{-2/3} \end{aligned} \tag{4.40}$$

Equation 4.39 ensures a bounded I', so, according to Thm. 4.8, the singularity must be a collision. It remains to show that a collision requires U to approach infinity in the indicated manner. This Eq. 4.40 estimate captures the behavior of colliding particles.

To see how Thm. 4.10 requires Eq. 4.37 to hold for all collisions, notice that $J^{1/2}$ imposes an upper bound $r_{j,k} \le A_1 (t^* - t)^{2/3}$ on all colliding particles while U^{-1} imposes a similar lower bound. The asymptotic form of Eq. 4.37 is obtained later *for some collisions*. As described later, we do not know whether this equation holds for all collisions because we do not know whether there are a finite number of central configurations. As such, the following essentially captures much of what is known for *all* possible collisions.

Corollary 4.1 *If there is a collision at time $t - t^*$ for the Newtonian N-body problem, then for each pair of colliding particles, j, k there are positive constants $A_{j,k}$, $B_{j,k}$ so that sufficiently close to the time of collision we have*

$$A_{j,k}(t^* - t)^{2/3} \le r_{j,k} \le B_{j,k}(t^* - t)^{2/3}.$$

4.5.2 Tauberian Theorems

This analysis of collisions relies on the so-called elementary Tauberian theorems. This topic was extensively advanced in the early twentieth century by G. Hardy and J. E. Littlewood. As a quick way to introduce what they do and why they are valuable, notice that it is permissible to integrate asymptotic relations and inequalities. For instance, knowing that

$$f'(t) \sim t \text{ as } t \to \infty$$

means that

$$f \sim \frac{1}{2} t^2 \text{ as } t \to \infty.$$

The converse, however, need not be true. To illustrate, notice that

$$f(t) = \frac{1}{2}t^2 + \cos(t^4) \sim \frac{1}{2}t^2 \text{ as } t \to \infty.$$

Rather than having $f' \sim t$, its derivative

$$f'(t) = t + 4t^3 \sin(t^4)$$

experiences a more radical growth rate—t^3 rather than t^2—than even f. The explanation is that inequalities cannot be differentiated, and asymptotic relationships are special forms of inequalities.

There are many situations—such as suggested in Sect. 2.2.1 (page 36)—where obtain new results will follow if we can differentiate an asymptotic relationship. The challenge is to find appropriate conditions on a function f to allow inequalities and asymptotic relations to be differentiated.

The above example suggests what is needed. Namely, the radical f' behavior reflects the rapid oscillations caused by the $\cos(t^4)$ term in f. Perhaps an asymptotic relationship can be differentiated if a function does not admit rapid oscillations. But if oscillations cause unexpected behavior in f', they should cause even greater problems with f''. So if f'' roughly "behaves somewhat as we might hope from the asymptotic relationship on f," then maybe the same tame behavior occurs with f'.

This observation captures the spirit of the Hardy and Littlewood results. The following illustrates a typical conclusion. (See Widder [111].)

Theorem 4.11 *For $f(t) \in C^2$, (i.e., f is twice differentiable and the derivatives are continuous), and if*

$$f(t) \sim At^\alpha, \alpha \neq 0, \text{ as } t \to \infty \text{ (or as } t \to 0)$$

and if

$$f'' > -Bt^{\alpha-2},$$

then

$$f'(t) \sim \alpha A t^{\alpha-1} \text{ as } t \to \infty \text{ (or as } t \to 0).$$

A similar relationship holds after replacing both "$\sim$" asymptotic symbols with the appropriate "$O(t^\beta)$" where $f(t) = O(t^\beta)$ means that $|\frac{f(t)}{t^\beta}|$ is bounded as $t \to \infty$ (or as $t \to 0$).

To develop intuition about these analytic results, it is worth proving Thm. 4.11 for the special and easy case where $f = O(t)$ as $t \to \infty$: the goal is to show that if $f'' > -B/t$, then $f' = O(1)$ as $t \to \infty$.

The assumption on f means that $f(2t) - f(t) = O(2t) - O(t) = O(t)$. Using Taylor's Theorem to expand f about t, the definition of $O(t)$ ensures there is a positive constant A so that, eventually,

$$At \geq f(2t) - f(t) = tf'(t) + \frac{t^2}{2} f''(\xi) \text{ where } t < \xi < 2t.$$

But $f''(t) > -Bt^{-1}$, so this equation can be expressed as the inequality

$$At \geq tf'(t) - \frac{t^2}{2} \frac{B}{\xi} \geq tf'(t) - \frac{t^2}{2} \frac{B}{t}.$$

Thus, eventually, $A + B \geq f'(t)$. To prove that $f'(t)$ eventually is bounded below, use Taylor's Theorem with $f(t) - f(2t)$ and expand f about $2t$.

Tauberian theorems and the N-body problem

To motivate why these theorems are useful in the study of N-body systems, the so-called "virial theorems" of mechanics describe the average behavior of the velocity and self-potential. These theorems, which are used by astronomers for identifying mass and other properties of systems, were introduced in Sect. 2.3.1 (page 47). A typical result asserts

Proposition 4.2 *If all $r_{j,k}$ are bounded above as $t \to \infty$ (denoted as $r_{j,k} = O(1)$ as $t \to \infty$), and all velocities are bounded above, then*

$$2 \lim_{t \to \infty} \frac{1}{t} \int_0^t T(s) \, ds = \lim_{t \to \infty} \int_0^t U(s) \, ds \quad \text{as } t \to \infty. \tag{4.41}$$

Pollard [63] significantly extended these conditions by use of elementary Tauberian theorems: his result totally drops the assumption on the velocities—assumptions that probably could never be verified—and relaxes the severe restriction that of bounded motion. Instead of bounded motion, all Pollard needs is that $\frac{r_{i,j}}{t} \to 0$ for each mutual distance; e.g., his result hold for unbounded motion where, say, $r_{i,j} < t/\ln(t)$.

The following minor extension of Pollard's result extends his conclusion to wider settings such as $r_{i,j} \sim A_{i,j}t$. Pollard's result [63] is where $A = 0$ with the $I \sim At^2$ assumption.

Theorem 4.12 *If either $I = O(t^2)$ or $I \sim At^2$ as $t \to \infty$, then*

$$\lim_{t \to \infty} \frac{1}{t} \int_0^t T(s) \, ds = O(1), \quad \lim_{t \to \infty} \frac{1}{t} \int_0^t U(s) \, ds = O(1).$$

Indeed, if $I \sim At^2$, then

$$\lim_{t\to\infty} \frac{1}{t} \int_0^t T(s)\, ds \to 2A - h, \quad \lim_{t\to\infty} \frac{1}{t} \int_0^t U(s)\, ds \to 2(A - h).$$

In particular, if $I = O(1)$, then $A = 0$, $h < 0$ and

$$\lim_{t\to\infty} \frac{1}{t} \int_0^t T(s)\, ds \to |h|, \quad \lim_{t\to\infty} \frac{1}{t} \int_0^t U(s)\, ds \to 2|h|.$$

More generally, let $I \sim At^\alpha$ as $t \to \infty$. If $\alpha > 2$, it follows that

$$\lim_{t\to\infty} \frac{1}{t^{\alpha-1}} \int_0^t T(s)\, ds = \lim_{t\to\infty} \frac{1}{t^{\alpha-1}} \int_0^t U(s)\, ds \to \alpha A.$$

If $\alpha < 2$ and $h \geq 0$, then it must be that $h = 0$ and

$$\lim_{t\to\infty} \frac{1}{t} \int_0^t T(s)\, ds = \lim_{t\to\infty} \frac{1}{t} \int_0^t U(s)\, ds = 0.$$

Proof: The proof follows from Thm. 4.11 where the Tauberian condition $f''(t) = I''(t) > -Bt^{\alpha-1}$ is supplied by the Lagrange-Jacobi equation Eq. 4.23, which ensures that $I'' = U + 2h > 2h$. If $\alpha \geq 2$, the Tauberian condition is automatically satisfied for all N-body problems; if $\alpha < 2$, the Lagrange-Jacobi condition satisfies the Tauberian condition for $h \geq 0$.

So either $I' = O(t^{\alpha-1})$ or $I' \sim \alpha A t^{\alpha-1}$ as $t \to \infty$ depending, respectively, on whether $I = O(t^\alpha)$ or $I \sim At^\alpha$ as $t \to \infty$. To complete the proof integrate the Lagrange-Jacobi equation to obtain

$$I'(t) - I'(0) = \int_0^t T(s)\, ds + ht = \int_0^t U(s)\, ds + 2ht.$$

If $\alpha < 2$, the added $h \geq 0$ assumption satisfies the Tauberian condition, so $I' \sim \alpha A t^{\alpha-1}$. As $\alpha - 1 < 1$, the assertion about the time averages follows.

The comment that $I = O(1)$ requires $h < 0$ follows from Prop. 4.1. If $h > 0$, we have from the Lagrange-Jacobi equation that $I'' \geq 2h$. Integrating twice, $I \geq ht^2 + I'(0)t + I(0)$ shows that $h > 0$ requires $\alpha \geq 2$, so the assertion about $\alpha < 2$ and $h = 0$ follows. $\square$

Elementary Tauberian theorems probably were first used in celestial mechanics to study collisions: they play a pivotal role in this analysis. To expand on a story[16] told in Winter's book [112], when R. Boas was starting his

---

[16]Ralph Boas was the Chair of the Northwestern Mathematics Department when he recruited me as an Asst. Professor. I heard this story from Boas before finding it in Winter's book.

mathematical career, Wintner suggested that a Tauberian theorem probably is embedded in Sundman's arguments [106, 107] about the behavior of binary collisions. Wintner's prediction was correct, and Boas [5] published the theorem. Boas' theorem provided the appropriate insight to prove my result about the general N-body collisions: a result that Pollard and I sharpened to the version in [64]. It is delightful how extensions of Sundman's insight completed the circle to return to assist in the study of collisions.

Nonlinear Tauberian conditions

Our purposes require a sharper, nonlinear version of Thm. 4.11. Rather than the inequalities $f'' < Bt^{\alpha-2}$ or $f'' > -Bt^{\alpha-2}$, we need to impose a bound on f'' in terms of powers of f'. A result in this direction was developed by Pollard [62] for his analysis of expanding gravitational systems (Pollard [61]): generalized versions are in Saari [73, 81].

Essentially, the generalized non-linear condition allows the $t^{\alpha-2}$ bound on f'' to be replaced by the appropriate combination of powers of f' and t^γ that would capture the traditional bound. For instance, *if $f \sim t^{\alpha-1}$,* then the bound $f'' < |f'|^\beta t^\gamma$ would replace the traditional $f'' < At^{\alpha-2}$ if β satisfies $[t^{\alpha-1}]^\beta t^\gamma = t^{\alpha-2}$. The following theorem is stated for $f \sim At^\alpha$ but it holds for $f = O(t^\alpha)$ and even for series where the derivative condition is replaced with a condition expressed in terms of differences.

Theorem 4.13 *(Saari [73, 81]) For $f(t) \in C^2$, if*

$$f(t) \sim At^\alpha, \alpha \neq 0, 1 \ \textit{as } t \to \infty \ \textit{(or as } t \to 0)$$

and if

$$f'' < B|f'|^{(\alpha-\gamma-2)/(\alpha-1)}t^\gamma,$$

then

$$f'(t) \sim \alpha At^{\alpha-1} \ \textit{as } t \to \infty \ \textit{(or as } t \to 0).$$

An application

Nonlinear Tauberian conditions are not standard tools, so let me illustrate them with another application. Suppose in the study of singularities, where $U \to \infty$, it turns out that

$$\int_t^a U(\tau)\,d\tau \sim At^{1/3} \ \text{as } t \to 0 \tag{4.42}$$

where A is a positive constant. We want to derive the sharper result

$$U \sim \frac{1}{3} A t^{-2/3} \text{ as } t \to 0, \tag{4.43}$$

which would require differentiating the Eq. 4.42 asymptotic relationship.

To justify differentiating this asymptotic relationship by using Thm. 4.13 where $\gamma = 0$ and $\alpha = \frac{1}{3}$ so $\beta = \frac{5}{2}$, we need to prove that

$$|U'| \leq A U^{5/2}. \tag{4.44}$$

Interestingly, Eq. 4.44 has been known to be true since the work of Sundman. To see why, recall from Cor. 2.1 (page 42) that positive constants A_1, A_2 can be found so that

$$\frac{A_1}{r_{min}} < U < \frac{A_2}{r_{min}}. \tag{4.45}$$

To verify Eq. 4.44, observe that by differentiating U and using Eq. 4.45 we obtain the string of inequalities

$$|U'| \leq \sum_{j \neq k} \frac{m_j m_k}{r_{j,k}^2} |r'_{j,k}| \leq \frac{A}{r_{min}^2} \sum_{j \neq k} m_j m_k |\mathbf{v}_j - \mathbf{v}_k|$$
$$\leq A U^2 \sum_{j \neq k} m_j m_k (|\mathbf{v}_j| + |\mathbf{v}_k|),$$

or

$$|U'| \leq A U^2 T^{1/2} \tag{4.46}$$

where the constant A represents different values with each usage and the last inequality uses the fact that there is some constant so that $|\mathbf{v}_k| \leq A\sqrt{T}$. So, if $h \leq 0$ (where h is the constant of energy), then $U \geq T$ and Eq. 4.44 holds. If $h > 0$ and if U is bounded away from zero (recall, as a collision is approached, $U \to \infty$ so this condition always is satisfied for singularities), then there is some constant so that $T < AU$, and the assertion holds.

Now that the Tauberian condition Eq. 4.44 has been verified, it follows from the nonlinear Tauberian theorem Thm. 4.13 that

Corollary 4.2 *If*

$$\int_t^a U(\tau)\, d\tau \sim A t^{1/3} \text{ and } U \to \infty \text{ as } t \to 0,$$

then

$$U \sim \frac{1}{3} A t^{-2/3} \text{ as } t \to 0.$$

4.5.3 Proof of the theorem

Armed with appropriate Tauberian theorems, we can prove Thm. 4.10. Use the change of variables $\tau = t^* - t$ to assume that the singularity occurs at $t = 0$ and that it is being approached through positive values. This assumption holds throughout the proof.

Before starting the proof, let me provide a map of what will be done. The first two subsections involve house-keeping: they derive appropriate and useful relationships between J and I. Next I show that J satisfies a version of the Lagrange-Jacobi equation: this is the equation that makes the polar moment of inertia I such an important N-body tool.

The real work starts around page 182 where I analyze the somewhat unexpected relationship $J'/J^{1/4}$. This is a crucial step, so let me motivate why we should expect this term to arise. The argument uses the reality that a surprising amount of what we do for the N-body problem involves finding appropriate ways to mimic what is done with the central force problem.

Recall from page 145 that with a central force problem $x'' = -1/x^2$ and the energy integral

$$\frac{1}{2} x'^2 = \frac{1}{x} + h,$$

the collision behavior is found by integrating the asymptotic relationship

$$x^{1/2} x' \sim A > 0 \text{ as } t \to 0.$$

To transfer the essence of this approach to the N-body collision problem, we need to translate the above asymptotic expression into recognizable N-body terms: this new expression involves J. To do so, notice that in the collinear central force setting, J becomes $J - x^2$ so $J' - 2xx'$. Consequently, $x^{1/2} = J^{1/4}$ and $x' = J'/2x = J'/2\sqrt{J}$. This means that

$$x^{1/2} x' = J^{1/4} \frac{J'}{2J^{1/2}} = \frac{J'}{2J^{1/4}}.$$

So, if there is any hope for notions from the central force analysis to extend to the N-body problem, we must

- find an argument replacing the asymptotic relationship $x^{1/2} x' \sim A$ as $t \to 0$ with the asymptotic expression $J'/J^{1/4} \sim A > 0$,

- and follow the lead of the central force problem by integrating this expression.

After carrying out this argument, the above Tauberian theorems complete our work.

Relating J and I

We must show that if the singularity is a collision, then J, J', and U have the indicated asymptotic expansions but now expressed as $t \to 0+$. The first step is to relate J and I. This is done by defining $\mathbf{c}_s$ as the center of mass of the particles with indices in G_s. Namely,

$$M_s \mathbf{c}_s = \sum_{k \in G_s} m_k \mathbf{r}_k, \quad M_s = \sum_{k \in G_s} m_k. \qquad (4.47)$$

For a fixed $k \in G_s$, express the equations of motion (Eq. 2.1, page 32) into a sum of terms that are in G_s and those that are not: this leads to

$$m_k \mathbf{r}_k'' = \sum_{j \neq k, j \in G_s} \frac{m_j m_k (\mathbf{r}_j - \mathbf{r}_k)}{r_{j,k}^3} + \sum_{j \notin G_s} \frac{m_j m_k (\mathbf{r}_j - \mathbf{r}_k)}{r_{j,k}^3}.$$

The second sum is bounded because for each $r_{j,k}$ term, the $\mathbf{r}_j$ and $\mathbf{r}_k$ are going to different limits (as $t \to 0$). Consequently we have

$$m_k \mathbf{r}_k'' = \sum_{j \neq k, j \in G_s} \frac{m_j m_k (\mathbf{r}_j - \mathbf{r}_k)}{r_{j,k}^3} + O(1) \text{ as } t \to 0+.$$

Applying these estimates into $\mathbf{c}_s'' = M_s^{-1} \sum_{k \in G_s} m_k \mathbf{r}_k''$ (Eq. 4.47) and using the anti-symmetry of the terms in the resulting double summation (as used to derive the center of mass integrals, page 33), we have that

$$\mathbf{c}_s'' = O(1) \text{ as } t \to 0+. \qquad (4.48)$$

We now can derive the desired relationship

$$J'' = I'' + O(1) \text{ as } t \to 0+. \qquad (4.49)$$

To do so, expanding the terms of J in Eq. 4.38 leads to

$$J = I - \sum_{s=1}^{q} M_s \mathbf{c}_s \cdot \mathbf{L}_s + \frac{1}{2} \sum_{s=1}^{q} M_s \mathbf{L}_s^2, \qquad (4.50)$$

where $\mathbf{L}_s$ represents the common $\mathbf{L}_j$ limit in G_s, so that

$$J'' = I'' - \sum_{s=1}^{q} M_s \mathbf{c}_s'' \cdot \mathbf{L}_s.$$

The desired Eq. 4.49 follows immediately from the last expression and Eq. 4.48.

The assumption that a singularity occurs as $t \to 0+$ requires $U \to \infty$ as $t \to 0+$, so $I'' \to \infty$ as $t \to 0+$. According to Eq. 4.49, J'' also approaches infinity as $t \to 0+$, so function J becomes concave up. By the definition of J, it follows that J never can be negative and $J(0+) = 0$.

These conditions prohibit J from approaching 0 through negative values, so they require J' to approach a non-negative limit as $t \to 0+$. This statement will be needed later on, so the assertion is recorded as

$$J' \to l \geq 0 \quad \text{as } t \to 0+. \tag{4.51}$$

The same argument along with the convexity (and monotonic growth of J' required by $J'' > 0$) means that $J' \geq 0$. This is because if J' were negative, it would require $J \to 0$ from negative values that is impossible.

Two useful inequalities

At this stage, we establish the useful inequalities

$$J'^2 \leq 4JT \tag{4.52}$$

and

$$J'' \geq \frac{A}{\sqrt{J}} \quad \text{as } t \to 0+. \tag{4.53}$$

Equation 4.52 follows by differentiating J to obtain

$$J' = \sum_{s=1}^{q} \sum_{k \in G_s} \sqrt{m_k}(\mathbf{r}_k - \mathbf{L}_k) \cdot \sqrt{m_k}\mathbf{v}_k,$$

which, by the Cauchy inequality, defines

$$J'^2 \leq [\sum_{s=1}^{q} \sum_{k \in G_s} m_k(\mathbf{r}_k - \mathbf{L}_k)^2][\sum_{s=1}^{q} \sum_{k \in G_s} m_k \mathbf{v}_k^2] = 4JT.$$

To derive Eq. 4.53, notice that by assuming there is a collision, at least two particles, say m_1 and m_2, approach the same limit $\mathbf{L}_1 = \mathbf{L}_2$. One of these two masses has the smaller value; assume it is m_1. This means that

$$\begin{aligned} J &\geq \tfrac{1}{2}[m_1(\mathbf{r}_1 - \mathbf{L}_1)^2 + m_2(\mathbf{r}_2 - \mathbf{L}_1)^2] \\ &\geq \tfrac{m_1}{2}[(\mathbf{r}_1 - \mathbf{L}_1)^2 + (\mathbf{r}_2 - \mathbf{L}_1)^2] \\ &\geq \tfrac{m_1}{4}[(\mathbf{r}_1 - \mathbf{r}_2]^2 = \tfrac{m_1}{4}r_{1,2}^2. \end{aligned}$$

Because $U \geq \frac{m_1 m_2}{r_{1,2}}$, the extreme ends of the above inequalities prove that $U \geq A/\sqrt{J}$. Substituting this inequality into $J'' = I'' + O(1) = U + O(1)$ (Eq. 4.49) along with $J \to 0$ leads to the desired Eq. 4.53 for some positive constant A.

The key step of the proof

If Thm. 4.10 is correct, then $J'/J^{1/4} \sim \frac{4}{3}At^{1/3}/[At^{4/3}]^{1/4}$ would approach a positive limit as $t \to 0+$. Indeed, the earlier comparison of this collision problem with the central force setting identifies the key step of the proof as showing that $J'/J^{1/4}$ approaches a positive limit. To prove this assertion, differentiate $J'/J^{1/4}$, and then integrate back again.

Differentiating and using $I'' = J'' + O(1)$ provides the expression

$$4\frac{d}{dt}\left(\frac{J'}{J^{1/4}}\right) = \frac{4JJ'' - J'^2}{J^{5/4}} = \frac{4JI'' - J'^2}{J^{5/4}} + O\left(\frac{1}{J^{1/4}}\right).$$

Using the Lagrange-Jacobi equation $I'' = U + 2h = T + h$ with the fact that $J'' = I'' + O(1)$ changes the relationship to

$$4\frac{d}{dt}\left(\frac{J'}{J^{1/4}}\right) = \frac{4JT - J'^2}{J^{5/4}} + O\left(\frac{1}{J^{1/4}}\right).$$

By integrating from t to $a > 0$, we have

$$4\left[\frac{J'}{J^{1/4}}(a) - \frac{J'}{J^{1/4}}(t)\right] = \int_t^a \frac{4JT - J'^2}{J^{5/4}}d\tau + O(1)\int_t^a \frac{d\tau}{J^{1/4}}. \qquad (4.54)$$

A way to show that $J'/J^{1/4}$ approaches a limit is to prove that both integrals on the right-hand side are convergent.

Combining the fact that J' remains positive and approaches a non-negative limit (Eq. 4.51) with the inequality from Eq. 4.52, we have that

$$J'J'' \geq \frac{AJ'}{J^{1/2}}$$

for t sufficiently close to 0. Integrating and using $J(0) = 0$ yields

$$J'^2 \geq C\sqrt{J}$$

for some positive constant C. In other words,

$$\frac{J'}{J^{1/4}} \geq \sqrt{C}, \text{ or, by integrating } J^{3/4} \geq \frac{3}{4}\sqrt{C}t. \qquad (4.55)$$

This lower bound for J ensures the convergence of

$$\int_{o+}^a \frac{d\tau}{J^{1/4}}.$$

To prove that the remaining integral in Eq. 4.54 converges, notice that because $J'/J^{1/4}$ is bounded away from zero (Eq. 4.55), Eq. 4.54 becomes the inequality

$$\int_t^a \frac{4JT - J'^2}{J^{5/4}} \, d\tau \le 4\frac{J'}{J^{1/4}}(a) + O(1).$$

Because Eq. 4.52 ensures that the integrand is non-negative, the integral is monotonically non-decreasing as $t \to 0$. But the integral is bounded, so

$$\int_{0+}^a \frac{4JT - J'^2}{J^{5/4}} \, d\tau$$

is convergent. Returning to Eq. 4.54 and letting $a = t_2$, $t = t_1$, we obtain

$$4[\frac{J'}{J^{1/4}}(t_2) - \frac{J'}{J^{1/4}}(t_1)] = \int_{t_1}^{t_2} \frac{4JT - J'^2}{J^{5/4}} d\tau + O(1) \int_{t_1}^{t_2} \frac{d\tau}{J^{1/4}}.$$

Because the right-hand side approaches zero, so does the left-hand side, and this establishes that

$$\frac{J'}{J^{1/4}} \to b > 0 \text{ as } t \to 0 + . \tag{4.56}$$

Completing the proof

By integrating this asymptotic relationship and using $J(0) - 0$, we have that $J^{3/4} \sim bt$, or the desired expression

$$J \sim At^{4/3} \text{ as } t \to 0 + .$$

The asymptotic relationship for J' follows if we can differentiate the relationship for J. According to Thm. 4.11, this can be done if $J'' \ge -Bt^{-2/3}$ for t sufficiently close to 0. But, according to Eq. 4.49, we have that $J'' = U + O(1)$, which means that $J'' > 0$ near zero. Thus, the Tauberian Thm. 4.11 ensures the expression

$$J' \sim \frac{4}{3}At^{1/3} \text{ as } t \to 0 + .$$

To complete the proof, by integrating $J'' = I'' + O(1) = U + O(1)$ from $a > 0$ to t we have that

$$J'(t) - J'(a) = \int_a^t U \, d\tau + O(t - a).$$

Letting $a \to 0$ and using the asymptotic bound for J' we have that

$$\int_{0+}^{t} U(\tau)\, d\tau \sim \frac{4}{3} A t^{1/3} \text{ as } t \to 0+ \, .$$

The final asymptotic conclusion

$$U \sim \frac{4}{9} \frac{A}{t^{2/3}} \text{ as } t \to 0+$$

now follows from the Tauberian theorem as reflected by Cor. 4.2. This completes the proof of Thm. 4.10. $\square$

4.6   Sharper asymptotic results

For each colliding pair $\mathbf{r}_i$ and $\mathbf{r}_j$, Cor. 4.1 (page 173) ensures there exist positive constants $A_{i,j}$ and $B_{i,j}$ so that, eventually,

$$B_{i,j}(t^* - t)^{2/3} \le r_{i,j}(t) \le A_{i,j}(t^* - t)^{2/3}. \tag{4.57}$$

It remains to improve $\mathbf{r}_j = \mathbf{L}_j + O((t^* - t)^{2/3})$ to the promised sharper asymptotic result that the particles approach a central configuration.

To do so, again let G_α be all indices with a common $\mathbf{L}_j$ as $t \to t^*$: some G_α sets are singletons (where a particle avoids colliding), while others have at least two indices. For convenience, let $t^* = 0$ and approach the collision through positive t values. Defining $\mathbf{L} = (\mathbf{L}_1, \dots, \mathbf{L}_N)$ and using the system coordinates of Chap. 2, we have as $t \to 0$ that $\mathbf{R} \to \mathbf{L}$.

For $i \in G_\alpha$, where G_α has at least two indices (it describes a collision), let $\tilde{\mathbf{R}}_i = (\mathbf{r}_i - \mathbf{L}_i)/t^{2/3}$. (So, the tilde represents scaled vectors.) In system coordinates, this defines $\tilde{\mathbf{R}} = (\mathbf{R} - \mathbf{L})/t^{2/3}$.

By differentiating $\mathbf{r}_i = \tilde{\mathbf{R}}_i t^{2/3} + \mathbf{L}_i$ twice, substituting terms into the equations of motion, and then multiplying both sides of the equations by $t^{4/3}$, the equations of motion in terms of $\mathbf{R}_i$ become

$$t^2 \tilde{\mathbf{R}}_i'' + \frac{4}{3} t \tilde{\mathbf{R}}_i' - \frac{2}{9} \tilde{\mathbf{R}}_i = \sum_{j \in G_\alpha} \frac{m_j(\tilde{\mathbf{R}}_j - \tilde{\mathbf{R}}_i)}{\tilde{R}_{i,j}^3} \\ + t^{4/3} \sum_{j \notin G_\alpha} \frac{m_j[(\tilde{\mathbf{R}}_j - \tilde{\mathbf{R}}_i)t^{2/3} + (\mathbf{L}_j - \mathbf{L}_i)]}{|(\tilde{\mathbf{R}}_j - \tilde{\mathbf{R}}_i)t^{2/3} + (\mathbf{L}_j - \mathbf{L}_i)|^3} \, . \tag{4.58}$$

Notice how Eq. 4.58 identifies how to prove that colliding particles must satisfy the central configuration equations. As $t \to 0$, the second summation on the right-hand side approaches zero, so the central configuration assertion follows by proving that $t^2 \tilde{\mathbf{R}}_i''$, $t \tilde{\mathbf{R}}_i' \to \mathbf{0}$ as $t \to 0$.

Theorem 4.14 *(Saari [86]) As a collision is approached, each particle in a G_α set, $|G_\alpha| \geq 2$, satisfies*

$$\frac{2}{9}\tilde{\mathbf{R}}_i + \sum_{j \in G_\alpha, j \neq i} \frac{m_j(\tilde{\mathbf{R}}_j - \tilde{\mathbf{R}}_i)}{\tilde{R}_{i,j}^3} \to \mathbf{0} \text{ for all } i \in G_\alpha. \tag{4.59}$$

Proof: To prove that $t\tilde{\mathbf{R}}_i' \to \mathbf{0}$, it suffices to show with system coordinates that $< \tilde{\mathbf{R}}', \tilde{\mathbf{R}}' >= o(t^2)$. But

$$\begin{aligned} < \tilde{\mathbf{R}}', \quad \tilde{\mathbf{R}}' >&=< \tfrac{d}{dt}\tfrac{\mathbf{R}-\mathbf{L}}{t^{2/3}}, \tfrac{d}{dt}\tfrac{\mathbf{R}-\mathbf{L}}{t^{2/3}} > \\ &=< -\tfrac{2}{3}(\mathbf{R}-\mathbf{L})t^{-5/3} + \mathbf{R}'t^{2/3}, -\tfrac{2}{3}(\mathbf{R}-\mathbf{L})t^{-5/3} + \mathbf{R}'t^{2/3} > \\ &= \tfrac{4}{9}t^{-10/3}(\mathbf{R}-\mathbf{L})^2 - \tfrac{4}{3}t^{\,2/3} < \mathbf{R}-\mathbf{L}, \mathbf{R}' > + t^{-4/3}\mathbf{R}'^2. \end{aligned} \tag{4.60}$$

The magnitude of the right-hand side of Eq. 4.60 is bounded by using Thm. 4.10. This is because $< \mathbf{R}-\mathbf{L}, \mathbf{R}-\mathbf{L} >= J$, while $< \mathbf{R}-\mathbf{L}, \mathbf{R}' >= \tfrac{1}{2}J'$, and $< \mathbf{R}', \mathbf{R}' >= 2U + 2h$. Substituting the three Thm. 4.10 asymptotic estimates into Eq. 4.60 leads to the desired conclusion $t^2 < \tilde{\mathbf{R}}', \tilde{\mathbf{R}}' >\to 0$.

To prove that $t^2\tilde{\mathbf{R}}'' \to \mathbf{0}$, use the Tauberian theorem Thm. 4.11 (page 174) asserting that if $f(t) = o(t^\alpha)$ and $f''(t) > Bt^{\alpha-2}$, then $f' = o(t^{\alpha-1})$. Our setting has that $\tilde{\mathbf{R}}' = o(t^{-1})$, so the conclusion follows if the components of $\tilde{\mathbf{R}}'''$ are bounded below by some multiple of t^{-3}.

To show that $t^3\tilde{\mathbf{R}}'''$ is bounded, we first show that $t^2\tilde{\mathbf{R}}''$ is bounded. According to Eq. 4.57, each $\tilde{R}_{i,j}$ is bounded away from zero, so the right-hand side of Eq. 4.58 is bounded. But as $t\tilde{\mathbf{R}}' \to \mathbf{0}$ and as each $\tilde{\mathbf{R}}_j$ must be bounded above (again from Eq. 4.57), it follows that $t^2\tilde{\mathbf{R}}''$ is bounded.

By differentiating both sides of Eq. 4.58, we have that

$$t^2\tilde{\mathbf{R}}_j''' = -2t\tilde{\mathbf{R}}_j'' - \frac{4}{3}(\tilde{\mathbf{R}}_j' + t\tilde{\mathbf{R}}_j'') + O(\tilde{\mathbf{R}}').$$

According to the derived estimates, each term on the right-hand side of this expression is of the order $O(1/t)$. Consequently, $t^3\tilde{\mathbf{R}}'''$ is bounded: the theorem is proved. $\square$

4.7 Spin, or no spin?

An objective of Thm. 4.14 is to sharpen the earlier $\mathbf{r}_j = \mathbf{L}_j + O(t^{2/3})$ to

$$\mathbf{r}_j = \mathbf{L}_j + \mathbf{A}_j t^{2/3} + o(t^{2/3}). \tag{4.61}$$

To do so takes more effort. The problem is similar to the one Painlevé confronted when showing that a collision occurs iff $\mathbf{R} \to \Delta$. Does $\mathbf{R}$ approach a specific point in Δ (a collision), or could $\mathbf{R}$ approach the set Δ but no specific point (a non-collision singularity)? Likewise, while $\tilde{\mathbf{R}}_\alpha$ must approach the algebraic variety defined by

$$\frac{2}{9}\tilde{\mathbf{R}}_i + \sum_{j \in G_\alpha, j \neq i} \frac{m_j(\tilde{\mathbf{R}}_j - \tilde{\mathbf{R}}_i)}{\tilde{R}_{i,j}^3} = \mathbf{0},$$

does it approach a specific point, or the set?

To answer this question, we need to consider how the equations for a central configuration could define a continuum.

1. Which central configuration? Suppose it eventually is proved that there exists a continuum of central configurations: the colliding particles might follow the manifold of possible choices without ever approaching a specific one. We can handle some of these settings with an additional assumption: the interested reader can check Saari [86].

2. More interestingly, the $SO(3)$ rotation of a central configuration $\mathbf{A}$ defines a manifold $\mathcal{M}_\mathbf{A}$. This leads to a problem raised by Painlevé and repeated by Wintner [112], which is resolved here: can the colliding particles spin infinitely often? Because we now know (Eq. 4.59) that $\tilde{\mathbf{R}} \to \mathcal{M}_\mathbf{A}$, interpret this question as asking whether $\tilde{\mathbf{R}}$ approaches a specific point in $\mathcal{M}_\mathbf{A}$. In simpler terms, can the orientation of the target central configuration keep changing? Proving this cannot happen helps to establish Eq. 4.61.

3. Beyond the Painlevé-Wintner question, even if the limiting central configuration has a fixed orientation, could the particles spin infinitely often while approaching it? This question refines Eq. 4.61 by seeking information about its $o(t^{2/3})$ term. Clark Robinson and I [71] proved that such spinning motion *can* occur with non-zero angular momentum for the N-body problem (for certain masses) as $t \to \infty$: the limiting collinear central configuration is on the z-axis (so its orientation is fixed), but the particles spin infinitely often about the axis while approaching the limiting behavior like a spinning figure skater pulling in her arms. Such behavior *cannot* happen with collisions.

Resolving 2, the Painlevé-Wintner issue, proves that Eq. 4.61 holds for all settings with a finite number of central configurations. A natural way to prove this result is to use system coordinates. Namely, as with the system velocity decomposition, divide

$$\tilde{\mathbf{R}}' = \tilde{\mathbf{W}}_{rot} + \tilde{\mathbf{W}}_{scal} + \tilde{\mathbf{W}}_{config}. \tag{4.62}$$

For the system defined by particles in G_α, use the α subscript; that is, $\tilde{\mathbf{R}}_\alpha$. I should use $< -, - >_\alpha$ and $\mathbf{E}_{j,\alpha}$, but as there should be no confusion, I drop the α subscript.

In Eq. 4.62, the $\tilde{\mathbf{W}}_{\alpha,config}$ component is what forces the particles to form the Eq. 4.59 central configuration (and determines whether there are a finite number of configurations). The $\tilde{\mathbf{W}}_{\alpha,scal}$ term determines the $\frac{2}{9}$ coefficient. The $\tilde{\mathbf{W}}_{\alpha,rot}$ term determines whether the particles spin as they approach a central configuration: indeed, its definition shows that $\tilde{\mathbf{W}}_{\alpha,rot}$ consists of all $\tilde{\mathbf{R}}'_\alpha$ components causing any spin in the system.

4.7.1   Using the angular momentum

Resolving the two different "spin issues" for complete collapse is easy. A N-body complete collapse can occur only if $\mathbf{c} = \mathbf{0}$ (Thm. 4.3 page 156). As shown in Chap. 2, $\mathbf{c} = \mathbf{0}$ requires $\mathbf{W}_{rot} \equiv \mathbf{0}$. A direct substitution proves that $\mathbf{W}_{rot} \equiv \mathbf{0}$ forces $\tilde{\mathbf{W}}_{rot} \equiv \mathbf{0}$; i.e., there is absolutely no rotation in the $\tilde{\mathbf{R}}'$ variable. Thus, spinning cannot occur.

For general collisions, part of the proof shows that the angular momentum of the colliding particles in G_α approaches zero fast enough so that $\mathbf{W}_{\alpha,rot}$ is integrable. By being integrable, the rotation cannot be infinite: this solves both the Painlevé-Wintner question and the extension introduced above. As described below, difficulties in this program arise when the limiting central configuration is collinear. Here we need more information about collisions. (The "non-spin" comments hold even if there is a continuum of central configurations.)

In stating the theorem, if $\tilde{\mathbf{R}}_\alpha = \mathbf{A}_\alpha$ represents a central configuration for the particles in G_α, then denote the $SO(3)$ orbit of $\mathbf{A}_\alpha$ by $\mathcal{M}_{\mathbf{A}_\alpha}$.

Theorem 4.15 *(Saari [86]) For a collision of particles in set G_α, not only does $\tilde{\mathbf{R}}_\alpha \to \mathcal{M}_{\mathbf{A}^*_\alpha}$ for some $\mathbf{A}^*_\alpha$, but there is a $\mathbf{A}_\alpha \in \mathcal{M}_{\mathbf{A}^*_\alpha}$ so that $\tilde{\mathbf{R}}_\alpha \to \mathbf{A}_\alpha$. Namely, for each $i \in G_\alpha$, there is $\mathbf{A}_i$ so that $(\mathbf{r}_i - \mathbf{L}_i)/t^{2/3} \to \mathbf{A}_i$. As the particles approach the limiting central configuration, they cannot rotate infinitely often about it. For a complete collapse problem, the colliding particles have no rotation whatsoever because $\tilde{\mathbf{W}}_{rot} \equiv \mathbf{0}$; for a general collision, $\mathbf{W}_{rot,\alpha} = O(t^{1/3})$.*

Not only is the rotational velocity integrable with a collision, but it rapidly approaches zero. This comment more dramatically captures the lack of spin.

Proof: The angular momentum for the colliding particles is used to determine the rotational velocity. Toward this end, let

$$\mathbf{c}_\alpha = \sum_{j \in G_\alpha} m_j \tilde{\mathbf{R}}_j \times \tilde{\mathbf{R}}_j'.$$

Mimicking the derivation of the angular momentum, take the cross product of $\tilde{\mathbf{R}}_i$ with both sides of Eq. 4.58, and sum over all $j \in G_\alpha$ to obtain $t^2 \mathbf{c}_\alpha' + \frac{4}{3} t \mathbf{c}_\alpha = O(t^{4/3})$, or (after dividing both sides by $t^{2/3}$)

$$\frac{d}{dt}(t^{4/3} \mathbf{c}_\alpha) = t^{4/3} \mathbf{c}_\alpha' + \frac{4}{3} t^{1/3} \mathbf{c}_\alpha = O(t^{2/3}).$$

Integrating this last expression, and then dividing by $t^{4/3}$, leads to $\mathbf{c}_\alpha = \mathbf{d} t^{-4/3} + O(t^{1/3})$ where $\mathbf{d}$ is a constant of integration. But each $\tilde{\mathbf{R}}_j$ is bounded, and, as established in the proof of Thm. 4.14, $\tilde{\mathbf{R}}_j' = o(t^{-1})$. As these estimates prove that $\mathbf{c}_\alpha = o(t^{-1})$, it must be that $\mathbf{d} = \mathbf{0}$ and $\mathbf{c}_\alpha = O(t^{1/3})$.

To find the behavior of $\tilde{\mathbf{W}}_{\alpha, rot}$, recall from Chap. 2 that a basis is given by $\{\mathbf{E}_j \times \tilde{\mathbf{R}}_\alpha\}$. As in Chap. 2, the $\tilde{\mathbf{W}}_{\alpha, rot}$ components are $[\mathbf{e}_j \cdot \mathbf{c}_\alpha]/(\mathbf{E}_j \times \tilde{\mathbf{R}}_\alpha)^2$. The numerator is of order $O(t^{1/3})$, which is integrable. Any problems, then, are caused by the denominator.

The denominator $< \mathbf{E}_j \times \tilde{\mathbf{R}}_\alpha, \mathbf{E}_j \times \tilde{\mathbf{R}}_\alpha >= \tilde{\mathbf{R}}_\alpha^2 - \tilde{\mathbf{R}}(j)_\alpha^2$ where $\tilde{\mathbf{R}}(j)_\alpha$ is the G_α system vector with the $\tilde{\mathbf{R}}_\alpha$ components in the $\mathbf{E}_j$ direction. (Horrible notation, but I am running out of symbols.) This means that problems arise when $\tilde{\mathbf{R}}_\alpha - \tilde{\mathbf{R}}(j)_\alpha \to \mathbf{0}$. For this expression to hold, the configuration must line up with the axis of rotation. Consequently this term is bounded away from zero for a coplanar problem (where the axis of rotation is $\mathbf{E}_3$, so the denominator is $\tilde{R}_\alpha^2$), or for any non-collinear central configuration. As the resulting velocity components are integrable, the conclusion follows.

It remains to show that $\tilde{\mathbf{W}}_{rot} \equiv \mathbf{0}$ for a complete collapse orbit. This follows by substituting terms into the standard angular momentum

$$\mathbf{0} = \sum_j m_j \mathbf{r}_j \times \mathbf{v}_j = \sum_j m_j \tilde{\mathbf{R}}_j t^{2/3} \times (\tilde{\mathbf{R}}_j t^{-1/3} + \tilde{\mathbf{R}}_j' t^{2/3}),$$

or $\mathbf{c}_\alpha = \sum_j m_j \tilde{\mathbf{R}}_j \times \tilde{\mathbf{R}}_j' = 0$. It now follows that $\tilde{\mathbf{W}}_{rot} \equiv \mathbf{0}$. $\square$

The denominator $< \mathbf{E}_j \times \tilde{\mathbf{R}}_\alpha, \mathbf{E}_j \times \tilde{\mathbf{R}}_\alpha >= \tilde{\mathbf{R}}_\alpha^2 - \tilde{\mathbf{R}}(j)_\alpha^2$ of the basis vectors for $\mathbf{W}_{rot,\alpha}$ is where the action is hiding. For spin to occur, the configuration is collinear and lines up with the angular momentum. Notice: it is not the orientation of the limiting configuration that can change (so the Painlevé-Wintner problem is answered), but whether the particles can spin about the limit. As shown next, they cannot.

4.7.2 The collinear case

The only setting not covered by the above proof, which involves a collinear configuration, is where detailed information about collision orbits is easy to find. I start with the complete collapse case.

Equation 4.58 is an Euler equation, so change it into an equation with constant coefficients. Namely, the $u = \ln(t)$ change of independent variables converts Eq. 4.58 to

$$\tilde{\mathbf{R}}'' + \frac{1}{3}\tilde{\mathbf{R}}' - \frac{2}{9}\tilde{\mathbf{R}} = \nabla_s U(\tilde{\mathbf{R}}), \tag{4.63}$$

or

$$\begin{aligned} \tilde{\mathbf{R}}' &= \tilde{\mathbf{V}} \\ \tilde{\mathbf{V}}' &= -\tfrac{1}{3}\tilde{\mathbf{V}} + \tfrac{2}{9}\tilde{\mathbf{R}} + \nabla_s U(\tilde{\mathbf{R}}), \end{aligned} \tag{4.64}$$

where the primes now represent differentiation with respect to u and where $t \to 0$, $t \to \infty$, correspond, respectively, to $u \to -\infty$ and $u \to \infty$.

Let $\tilde{\mathbf{R}} = \mathbf{A}$ represent a central configuration. The $SO(3)$ orbit of $\mathbf{A}$ defines a manifold, denoted by $\mathcal{M}_{\mathbf{A}}$, and all points on $\mathcal{M}_{\mathbf{A}}$ are central configurations. According to Thm. 4.14, $(\mathbf{A}, \tilde{\mathbf{V}} = \mathbf{0})$ is an equilibrium point for Eq. 4.64, so it is natural to compute its stable and unstable manifolds. The stable manifold is where solutions approach $(\mathbf{A}, \mathbf{0})$ as $u \to \infty$, or as $t \to \infty$: it represents expanding motions where the scaled system vector $\mathbf{R}/t^{2/3} = \tilde{\mathbf{R}}$ asymptotically approaches the central configuration $\mathbf{A}$ as $t \to \infty$. (This expansion is called "completely parabolic motion.") The unstable manifold, on the other hand, considers $\tilde{\mathbf{R}}$ solutions as $u \to -\infty$, or as $t \to 0$: the system is headed for a collision. We need to examine the unstable manifold to complete the analysis of the spin problem for collision orbits.

The standard approach (e.g., see Katok and Hasselblatt [29], or Robinson [68]), is to find the eigenvalues and eigenvectors for the linear approximation of right-hand side of Eq. 4.64: this structure governs the behavior of motion approaching the equilibrium. Represent the right-hand side of Eq. 4.64 as

$$\begin{pmatrix} 0 & E \\ B & -\frac{1}{3}E \end{pmatrix} \tag{4.65}$$

where E is the identity matrix and $B = D_{\mathbf{A}}(\nabla_s(\frac{1}{9}I(\tilde{\mathbf{R}}) + U(\tilde{\mathbf{R}})))$ is the differential of the gradient evaluated at $\mathbf{A}$. If λ is an eigenvalue for this system with eigenvector $(\mathbf{u}, \mathbf{v})$, we have that $\lambda\mathbf{u} = \mathbf{v}$ and $(\lambda + \frac{1}{3})\mathbf{v} = B\mathbf{u}$, or

$$\lambda(\lambda + \frac{1}{3}) = B\mathbf{u}.$$

So, if **u** is an eigenvector for B with eigenvalue μ, we have that

$$\lambda(\lambda + \frac{1}{3}) - \mu = 0.$$

In other words, the eigenstructure of B determines the eigenstructure for the full system.

The stable manifold, or expansion

To analyze the stable manifold, which characterizes the expanding parabolic motion as $t \to \infty$, the emphasis is placed on eigenvalues with negative real parts. But because

$$\lambda = \frac{-\frac{1}{3} \pm \sqrt{\frac{1}{9} + 4\mu}}{2}, \tag{4.66}$$

it follows that for all μ the real part of at least one of the corresponding pair of λ eigenvalues is negative. An interesting feature arises should the discriminant, $\frac{1}{9} + 4\mu$, be negative. Here, when $\mu < -\frac{1}{36}$, the corresponding λ pair has a negative real part and non-zero imaginary parts. These eigenvalues introduce solutions involving cosine and sine terms—possible rotations!

    This is how Robinson and I [71] proved for completely parabolic motion in the non-coplanar N-body problem that there are masses for which the orbits approach a limiting collinear configuration along the z-axis, but some particles spin infinitely often about this limit as $t \to \infty$. Our proof shows that certain mass values allowed $\mu < -\frac{1}{36}$: thus there are pairs of λ eigenvalues with negative real part and non-zero imaginary parts. (The eigenvectors are off the line defined by the collinear configuration.)

    The impact of these eigenvalues is that, as solutions approach **A**, certain lower order terms have cosine and sine components. By being lower order, rather than the *dominating term*, these values (which require a non-zero angular momentum) only slightly modify the asymptotic behavior; e.g., in t coordinates, the motion is dominated by the leading $t^{2/3}$ term so it does not have a significant rotating effect. The situation changes radically with the non-coplanar problem because, as we showed, the dominating $t^{2/3}$ terms can be centered along the z axis. The important consequence, for our purposes, is that the dominating terms in the x-y coordinates now have cosine and sine components! In this manner, we proved there exist solutions where particles have a slow $(t^{1/2}\cos(\ln(t^a)), t^{1/2}\sin(\ln(t^a)))$ rotation about the z-axis.

The unstable manifold, or collisions

The situation changes when studying collisions, or the unstable manifold. This manifold, characterized by eigenvalues with positive real parts, requires $\mu > 0$ (Eq. 4.66). Of particular importance for spin-question is that *it is impossible to have an eigenvalue with a non-zero imaginary part. Therefore, collisions never admit an infinite rotation or spin.*

I ignored $\mu = 0$, which causes one of the corresponding λ terms to be zero. This can occur because $\frac{1}{9}I + U$ is rotation invariant, so the kernel of B must include the tangent space $T_\mathbf{A}\mathcal{M}_\mathbf{A}$: this tangent space includes all directions corresponding to rigid body rotations of $\mathbf{A}$. Other than the natural invariances of $\frac{1}{9}I + U$, the eigenvector for any other $\mu = 0$ setting corresponds to a degenerate central configuration (Def 2.2, page 46). While such configurations exist, it is well known that the collinear central configurations are *non-degenerate*, so $\mu = 0$ does not cause problems.

The conclusion now follows from the notions of normally hyperbolic behavior: namely, all eigendirections off of $\mathcal{M}_\mathbf{A} \times \mathbf{0}$ correspond to non-zero eigenvalues. As this is true for all points, trajectories asymptotic to this set are in phase and each trajectory is asymptotic to a particular point. The unstable manifold is foliated into the unstable manifolds for each point in $\mathcal{M}_\mathbf{A} \times \mathbf{0}$. This completes the proof. $\square$

My proof for general collisions in [86] increased the dimension of the system by introducing a z variable to separate individual collisions into their own "total collapse" systems. The equations for G_α replace Eq. 4.64 with

$$
\begin{aligned}
z' &= \tfrac{2}{3}z \\
\tilde{\mathbf{R}}'_\alpha &= \tilde{\mathbf{V}}_\alpha \\
\tilde{\mathbf{V}}'_\alpha &= -\tfrac{1}{3}\tilde{\mathbf{V}}_\alpha + \tfrac{2}{9}\tilde{\mathbf{R}}_\alpha + \nabla_s U_\alpha(\tilde{\mathbf{R}}_\alpha) + z^2\mathbf{H}_\alpha,
\end{aligned}
\tag{4.67}
$$

where $U_\alpha = \sum_{i \neq j \in G_\alpha} \frac{m_i m_j}{r_{ij}}$ and $\sum_{k \notin G_\alpha} \frac{m_k[(\tilde{\mathbf{R}}_k - \tilde{\mathbf{R}}_i)z + (\mathbf{L}_k - \mathbf{L}_i)]}{|(\tilde{\mathbf{R}}_k - \tilde{\mathbf{R}}_i)z + (\mathbf{L}_k - \mathbf{L}_i)|^3}$ is the ith component of $\mathbf{H}_\alpha$. The advantage of doing so is that when $z = 0$, the simultaneous collisions separate into different "total collapse" problems, while the z variable only introduces another hyperbolic direction. This approach simplifies the transition between total collapse and simultaneous collisions. N-body solutions correspond to the special case where $z(1) = e^{\frac{2}{3}}$.

4.8   Manifolds defined by collisions

The structure used to handle the collinear case extends to all known kinds of collisions as well as parabolic behavior. The idea is straightforward: compute the eigenvalues for Eq. 4.67 and apply the appropriate stable manifold theorem. Sample results, from Saari [86] and Saari and Hulkower [92], are Thms. 4.16, 4.17.

Theorem 4.16 *If an orbit terminates in a total collapse where the limiting configuration is collinear, then the motion was confined to a straight line for all time. If it terminates in a coplanar central configurations, then the motion is coplanar for all time.*

The idea is that the dimension of the unstable manifold for colliding orbits does not change with the dimension of the space. The conclusion follows from uniqueness theorems. But, Thm. 4.16 does not hold for parabolic motion (as $u \to \infty$) as these manifold can change with the size of physical space. Indeed, notice from the above discussion and Eq. 4.66 how the unstable manifolds (collisions) tend to be more restrictive.

4.8.1   Structure of collision sets

What is the structure of sets of initial conditions leading to collisions? Siegel [100] showed for the three-body problem that they are smooth submanifolds. The same result does not quite hold for all collisions: problems arise with orbits tending to a degenerate central configuration as, here, the extra zero eigenvalue requires using the unstable-center manifold theorems.

Theorem 4.17 *The set of initial conditions tending toward a single, or simultaneous, collisions where all limiting central configurations are non-degenerate forms a set of C^∞ smooth submanifolds. If any of the limiting central configurations are degenerate, then the set of initial conditions is contained in such union of submanifolds.*

A word of caution: while Thm. 4.17 asserts that the collision bound set is determined by smooth submanifolds, it does *not* require the submanifolds to be imbedded: for some masses, they may not be. These results permit all sorts of corollaries for different values of N. For instance, the four-body problem allows a complete collapse, a triple collision, two simultaneous binary collisions, or a binary collision. The structure of the collision set, then, is a finite union of smooth lower-dimensional submanifolds—except for the degenerate four-body central configurations (as described on page 46) where the set is included in such a manifold.

Another analytical question concerning collisions, which is attributed to Siegel, is to determine the complex analytic classification of the collision singularities. By treating the independent variable t as a complex variable, solutions of the N-body problem become complex analytic functions of t. Sundman showed that a binary collision is an algebraic branch point. Furthermore, by use of Riemann surfaces, or a uniformizing change of variables, this type of singularity can be removed. On the other hand, Siegel [100] showed for most mass values that a total collapse of the three-body problem is a logarithmic branch point. What happens in general?

For collisions involving three or more particles, expect the singularity to be a logarithmic singularity. This is because for k-fold collisions with $k > 2$, the powers in the expansion can depend continuously on the masses, so the singularities tend to be essential singularities. But what about simultaneous binary collisions? The next theorem verifies an earlier assertion (Eq. 4.15, page 146) that a binary collision is an algebraic branch point by extending the assertion to any number of simultaneous binary collisions.

Theorem 4.18 *(Saari [86]) Any multiple collision that consists only of simultaneous binary collisions at time $t = 0$ is an algebraic branch point where the behavior of each particle can be expanded in a power series of $t^{1/3}$ that converges in a neighborhood of $t = 0$.*

4.8.2 Proof of Theorem 4.18

Proof of Thm. 4.18. The proof follows how we teach beginning students to solve differential equations with series solutions by collecting terms with like powers. The main difference involves a change of independent variables.

Assume that k binary collisions, with particles $\{1, 2\}, \{3, 4\}, \ldots, \{2k-1, 2k\}$, occur simultaneously at time $t = 0$: this defines the sets $G_j, j = 1, \ldots, k$. Using $s = t^{1/3}$ change of independent variable, Eq. 4.58 becomes

$$s^2\tilde{\mathbf{R}}_j'' + 2s\tilde{\mathbf{R}}_j' - 2\tilde{\mathbf{R}}_j = \frac{9}{m_j}\frac{\partial U_j}{\partial \tilde{\mathbf{R}}_j} + s^4\mathbf{g}_j(s, \tilde{\mathbf{R}}) \tag{4.68}$$

where $U_j = \frac{m_j m_{j+1}}{R_{j,j+1}}$ and the components of $\mathbf{g}_j(s, \tilde{\mathbf{R}})$ are analytic functions of the variables s and $\tilde{\mathbf{R}}$. (It is a multiple of the second summation of the right-hand side of Eq. 4.58 where t terms are replaced with the appropriate power of s and the primes represent differentiation with respect to s.) For non-colliding particles, the partial derivative term on the right-hand side of Eq. 4.68 is missing.

Next, substitute the series $\tilde{\mathbf{R}}_j = \sum_{\gamma=1}^{\infty} \mathbf{a}_{j,\gamma}s^\gamma$ into Eq. 4.68. As standard, use the analyticity of the terms on the right-hand side of the equation to find its power series expansion: terms on both sides are collected according to the s^γ powers. For $\tilde{\mathbf{R}}'_j$ and $\tilde{\mathbf{R}}''_j$ terms, formally differentiate the series and substitute into the equations. Let $\mathbf{a}_\gamma = (\mathbf{a}_{1,\gamma}, \ldots, a_{N,\gamma})$.

In collecting s^γ terms, notice that the s^4 multiple of $\mathbf{g}$ ensures that the formal recurrence relationship does not involve $\mathbf{g}$ for $\gamma = 0, \ldots, 3$. The $\gamma = 0$ term only involves $-2\tilde{\mathbf{R}}_j = \frac{9}{m_j} \frac{\partial U_j}{\partial \tilde{\mathbf{R}})_j}$, so this defines the central configuration. (This $\mathbf{a}_{j,0}$ term is zero for non-colliding particles.) For $\gamma = 1, \ldots$, the recurrence relationship assumes the form

$$[(\gamma - 1)(\gamma + 2)E - 9D(\nabla_j U_j)](\mathbf{a}_{2j-1,\gamma}, \mathbf{a}_{2j,\gamma}) = \mathbf{h}_j(\mathbf{a}_0, \ldots, \mathbf{a}_{\gamma-1}) \quad (4.69)$$

for the colliding particles and

$$(\gamma - 1)(\gamma + 2)E(\mathbf{a}_{j,\gamma}) = \mathbf{h}(\mathbf{a}_0, \ldots, \mathbf{a}_{\gamma-1}) \quad (4.70)$$

for the non-colliding particles where E is the identity matrix of the appropriate dimension and the $\mathbf{h}$ terms come from the $\mathbf{g}$.

For the non-colliding particles, it follows from Eq. 4.70 that $\mathbf{a}_{j,\gamma} = \mathbf{0}$ for $\gamma = 0, 2, 3$, while $\mathbf{a}_{j,1}$ is arbitrary— $\mathbf{a}_{j,1}$ corresponds to the limiting velocity of the non-colliding particle. The choices for $\mathbf{a}_{j,\gamma}$, $\gamma \geq 4$ are uniquely determined by Eq. 4.70. In general, $\mathbf{a}_{j,5} \neq \mathbf{0}$ because the corresponding $\mathbf{h}$ is the perturbation term $\mathbf{g}'(0, \mathbf{a}(0))$. This means that *even the non-colliding particles have an algebraic branch point at the collision.*

To determine the series coefficients for colliding particles, we must determine the properties of matrix $D(\nabla_j U_j)$. To find this matrix, we need the positions of the particles, which we assume to be particles 1 and 2. At $t = 0$, we have that $m_1\tilde{\mathbf{R}}_1 + m_2\tilde{\mathbf{R}}_2 = \mathbf{0}$, so $\tilde{R}_{1,2} = \tilde{R}_2((m_1 + m_2)/m_1)$. By combining the central configuration equations, we have

$$\frac{2}{9}(\tilde{\mathbf{R}}_2 - \tilde{\mathbf{R}}_1) = (m_1 + m_2)\frac{\tilde{\mathbf{R}}_2 - \tilde{\mathbf{R}}_1}{\tilde{R}_{1,2}^3},$$

or that $\tilde{R}_{1,2}^3 = 9(m_1 + m_2)/2$.

Using this value for $\tilde{R}_{1,2}$ and assuming that the configuration is on the x-axis, a direct computation shows that $9D(\nabla_j U_j)$ is the 6×6 matrix

$$\begin{pmatrix} C\{m_2\} & -C\{m_2\} \\ -C\{m_1\} & C\{m_1\} \end{pmatrix} \quad (4.71)$$

where $C\{m\}$ is the 3×3 diagonal matrix with $4m/(m_1 + m_2)$ in the $(1,1)$ position and $-2m/(m_1 + m_2)$ in the other two diagonal slots.

The three-dimensional kernel of this matrix is spanned by $(\mathbf{e}_j, \mathbf{e}_j)$ where $\mathbf{e}_j$ is the usual unit three-vector with unity in the jth component. This kernel reflects the translation invariance $U_1(\tilde{\mathbf{R}}_1, \tilde{\mathbf{R}}_2) = U_1(\tilde{\mathbf{R}}_1 + \mathbf{D}, \tilde{\mathbf{R}}_2 + \mathbf{D})$ for any three-vector $\mathbf{D}$. The non-zero eigenvalues are 4, with eigenvector $(1, 0, 0; m_1/m_2, 0, 0)$, and -2, which is accompanied by the eigenspace spanned by $(0, 1, 0; 0, -m_1/m_2, 0)$ and $(0, 0, 1; 0, 0, -m_1/m_2)$. With this information, we can return to computing the series representation for the colliding particles.

According to Eq. 4.68, the $\mathbf{h}$ function is zero for $\gamma = 1$. Thus, according to Eq. 4.69, the choices of $\mathbf{a}_{1,1}, \mathbf{a}_{2,1}$ come from the kernel of $D(\nabla_1 U_1)$. As the kernel corresponds to the translation of the binary system, they represent the final velocity of the center of mass of the colliding binary.

For $\gamma = 2$, $\mathbf{h}$ is given by $9D^2 \nabla_j U_j\{\mathbf{w}, \mathbf{w}\}$ where $\mathbf{w}$ is the six-vector selected for the $\gamma = 1$ term. But $\mathbf{w}$ corresponds to the translation invariance of U_j, so this term is equal to zero. Consequently, the $\gamma = 2$ value in Eq. 4.69, which creates a multiple of 4 for the E matrix, mandates that $\mathbf{a}_{1,2}, \mathbf{a}_{2,2}$ is a multiple of the above eigenvector for the eigenvalue 4.

For $\gamma > 2$, the matrix on the left-hand side of Eq. 4.69 is non-singular and the multiple of the matrix E clearly dominates the matrix sum, so the coefficients of the series can be uniquely determined. A counting argument establishes that the degrees of freedom in the choices of the $\mathbf{a}_\gamma$ correspond to the dimension of the unstable manifold for the multiple collision as determined earlier. Thus, all collision solutions are represented by one of these series solutions. Also, for $\gamma > 2$, the norm of the inverse of the matrix on the left-hand side goes to zero like $1/\gamma$, so a standard majorant argument (e.g., see Hille [26]) shows that all of these series solutions converge in some neighborhood of $s = 0$. This completes the proof. $\square$

4.9   Proof of the slowly varying assertion

I now prove Thm. 4.9, which asserts that *a singularity of the N-body problem at $t = 0$, where I is slowly varying as $t \to 0$, is caused by collisions.* As established on page 168, the von Zeipel theorem is a special case of this assertion because if $I = O(1)$ as $t \to 0$, I is slowly varying.

The proof exploits relationships among competing dynamics. Namely, with a non-collision singularity, infinitely often some particles are very close to others, forming natural groups that are separated by considerable distances. Then some particle must leave one group to visit another one: this causes a large change in some previously small $r_{i,j}$.

This larger $r_{i,j}$ value changes the location of the center of mass of the original group. How can this be? After all, the acceleration for the center of mass of any group is very small; it is of the order of the distance between groups squared. With a small acceleration and a large change in the position of the center of mass over a very short time interval, the velocity of the center of mass must be very large. The proof shows that if I is slowly varying, this velocity cannot be as large as it must be. The key estimate is captured by Fig. 4.5, which shows how the centers of mass of these groups must change: at some time t the second derivative of a measure of the size of these groups is small, and this bounds the velocities of the centers of mass. (This measure, "K," is introduced in the next paragraph.)

What complicates the analysis is that we have only rough qualitative knowledge about the dynamics. Thus the needed estimates for the behavior of the center of mass of any particular group are difficult to obtain. Therefore the scheme is to study and find estimates on the velocities for the centers of mass of all groups. As such, the discussion centers on $K = \frac{1}{2} \sum M_s \mathbf{c}_s^2$, which is a "moment of inertia" type measure for the centers of mass.

To start the proof, recall that with a singularity at $t = 0$, eventually I is concave up, and, as $t \to 0$, either I approaches a non-negative limit or infinity. Assuming that the singularity is not a collision rules out the possibility that $I \to 0$, which is a complete collapse and collision.

Counting arguments

As advertised, the first part of the proof shows that, infinitely often, a certain number of mutual distances are close to one another, while the others must be separated. The first lemma shows that a non-collision singularity must be accompanied by a rapid and radical oscillatory behavior.

Lemma 4.2 *With a non-collision singularity, there exist indices i, j so that*

$$\liminf_{t \to 0} \frac{r_{i,j}}{\sqrt{I}} \to 0, \quad \limsup_{t \to 0} \frac{r_{i,j}}{\sqrt{I}} > 0. \tag{4.72}$$

Proof: If the lemma were false, then, because N is finite, there is an $A > 0$ so that for each $r_{i,j}$, either $r_{i,j}/\sqrt{I} \to 0$, or, eventually, $r_{i,j}/\sqrt{I} > A > 0$. Use this fact to partition the indices into classes G_k, $k = 1, \dots, p < N$ where $i, j \in G_s$ iff $r_{i,j}/\sqrt{I} \to 0$ as $t \to 0$. Some classes may consist of a single entry.

Define the center of mass $\mathbf{c}_s$ for G_s as $M_s \mathbf{c}_s = \sum_{j \in G_s} m_j \mathbf{r}_j$, where $M_s = \sum_{i \in G_s} m_i$. Using the same argument as in Eq. 4.48 (page 180), it follows that $\mathbf{c}_s'' = O(1)$ as $t \to 0$. Consequently, both $\mathbf{c}_s'$ and $\mathbf{c}_s \to \mathbf{L}_s$ approach limits as $t \to 0$.

If $I = O(1)$, then the $r_{i,j}/\sqrt{I} \to 0$ condition implies that $r_{i,j} \to 0$, or that $\mathbf{r}_i, \mathbf{r}_j \to \mathbf{L}_s$ as $t \to 0$; that is, all particles in G_s approach the limiting position of the center of mass, so they all approach a collision. If $I \to \infty$, then consider the equality

$$\frac{1}{2} \sum_{s=1}^{p} \sum_{i \in G_s} m_j (\mathbf{r}_j - \mathbf{c}_s)^2 = I - \frac{1}{2} \sum_{s=1}^{p} M_s \mathbf{c}_s^2. \qquad (4.73)$$

As each $\mathbf{c}_s$ has a limit, the second summation on the right-hand side of Eq. 4.73 approaches a finite limit. By definition of the G_s sets and the requirement that $r_{i,j}/\sqrt{I} \to 0$, the summation on the left-hand side is $o(I)$. This leads to the contraction $I = o(I)$, which proves the lemma. $\sqcap$

Much of what follows is headed for comparisons of the form permitted by Eq. 4.73: the left-hand side measures changes among particles in each grouping, while the right-hand side is divided into changes of I and the centers of mass. But the small $\mathbf{c}_s''$ value determined by the particles *not* in G_s limits the behavior of $\sum M_s \mathbf{c}_s^2$. This restriction allows the assumed behavior of I to be compared with the required behavior of particles leaving a group, as captured by the left-hand side of Eq. 4.73.

Lemma 4.2 establishes that a non-collision singularity requires, infinitely often, oscillatory behavior where particles flirt with approaching one-another, but then separate. (It is easy to prove that the departing particle comes arbitrarily close to a particle from another group. If not, the departing particle moves off in essentially a straight line and would never return—contradicting the lemma.) The next lemma shows that infinitely often the particles can be partitioned into those that are reasonably close to one another, or far apart. This lemma asserts that there is a positive integer F so that, eventually, there always are at least F mutual distances larger than a particular bound.

Lemma 4.3 *There is a constant $A > 0$ so that for any $\epsilon > 0$, a sequence $\{t_i\}$, $t_i > t_{i+1} > \dots \to 0$, can be found so that for all $\{k, j\}$, either $r_{j,k}(t_i) < \epsilon \sqrt{I(t_i)}$ or $r_{j,k}(t_i) > 2A\sqrt{I(t_i)}$. There is a positive integer F so that, eventually, at least F mutual distances always satisfy $r_{j,k}(t) \geq 2A\sqrt{I(t)}$. (The same mutual distances need not always satisfy this inequality.)*

Proof: The first part of the argument establishes that, at certain times, a certain number of mutual distance are at least a fixed multiple of $\sqrt{I}$ apart. To prove this, let $\eta(\tau, d)$ be the number of mutual distances $r_{j,k}$ that, at $t = \tau$, satisfy $r_{j,k} \geq d\sqrt{I}$. According to Eq. 2.22 (page 42), there is a constant H depending only on the masses so that if $d < H$, then $1 \leq \eta(\tau, d) \leq \binom{N}{2}$.

To find the minimum number of separated distances that eventually holds for any τ, let $D(t,d) = \{\eta(\tau, d) \,|\, \tau \in (0,t),\, d < H\}$ be the number of separated mutual distances at each $t > 0$ as $t \to 0$. Because $d < H$ and the η values are positive integers, we have that $E(t,d)$, which is the greatest lower bound of $D(t,d)$, is a positive integer.

To find the largest d value that, eventually, always offers the same number of separated mutual distances, notice that $E(t_1, d) \geq E(t,d)$ for $t_1 < t$, and $\eta(\tau, d_1) \geq \eta(\tau, d)$ for $d_1 < d$ yield the statement

$$\text{if } d_1 \leq d,\, t_1 \leq t, \text{ then } E(t_1, d_1) \geq E(t,d). \tag{4.74}$$

If F is the least upper bound of $E(t,d)$ for $0 < t \leq 1$ and $0 < d \leq H$, then F is a positive integer. Thus there is a (t_0, d_0) so that $E(t_0, d_0) = F$. According to Eq. 4.74, $E(t,d) = F$ for $0 < t \leq t_0$, $0 < d \leq d_0$. To interpret these terms, t_0 defines "eventually"; it means for any $t < t_0$. Of particular importance, d_0 is the desired "optimal" value of d. Namely, as we started out to prove, at each instant of time in $(0, t_0]$, there always are at least F mutual distances where $r_{j,k} > d\sqrt{I}$ for $0 < d \leq d_0$.

This F value allows us to prove that, infinitely often, a specified number of mutual distances must be very small. To see this, let $2A$ be the minimum of d_0 and each of the $\limsup \frac{r_{j,k}}{\sqrt{I}}$ for which this limit superior is positive. Let positive $\epsilon < A$ be given. For each $t < t_0$, a positive value $t_i < t$ can be found so that precisely $[\binom{N}{2} - F]$ distances $r_{j,k}$ are bounded above by $\epsilon\sqrt{I}$. If this were not true, then we would have that $\eta(\tau, \epsilon) > F$ for $0 < \tau < t$, which would imply that $E(t, \epsilon) > F$, and that would contradict the definition of F and Eq. 4.74. Thus, the statement is proved.

So, at time t_i precisely F mutual distances are not bounded above by $\epsilon\sqrt{I}$. To add separation to these distances, recall that we have established for all $t < t_0$ that there are at least F mutual distances greater than $2A\sqrt{I}$. This proves the lemma. $\square$

4.9.1   Centers of mass

Use the above results to partition the N particles into sets of particles that are "close" to each other (as measured by $\sqrt{I}$). Again, some sets might be singletons. Thus, at least for short intervals of time, the "near one-another" particles define a cluster. Because the value of ϵ has not been specified, select δ so that

$$\delta < \min(m_0 A^2/8,\, A/2), \tag{4.75}$$

where m_0 is the smallest mass value, and let $\epsilon = (2\delta/M)^{1/2}$.

The first step is to establish the existence of short intervals of time when each mutual distance is either less than $A\sqrt{I}$, or greater than $2A\sqrt{I}$. That this must happen follows from the choice of $2A$ and Eq. 4.72: it cannot be that in some interval $[0, t]$, all mutual distances always are either bounded above by $A\sqrt{I}$ or always bounded below by $2A\sqrt{I}$. Thus, some particles must be commuting.

Start at time t_i described above where precisely $[\binom{N}{2} - F]$ distances $r_{j,k}$ are bounded above by $\epsilon\sqrt{I}$. Because F distances must always be bounded below by $2A\sqrt{I}$, it follows from Eq. 4.72 and the choice of A that some particle must break away from its group: its mutual distance with some particle in the original group eventually exceeds $2A\sqrt{I}$. Let $t_{i,1} < t_i < t_{i,2}$ be the first times on both sides of t_i when some distance equals $A\sqrt{I}$. (Different mutual distances, involving particles from different groups, may define the two endpoints of this interval.) In this small time interval ("small" because $t \to 0$), some mutual distance grows from being less than $\epsilon\sqrt{I}$ to equal $A\sqrt{I}$. This growth plays an important role in the proof.

The proof of the theorem involves changes in the centers of mass. As in the proof of Lemma 4.2, partition the indices into classes defined by "nearby" particles; i.e., $G_s = \{i, j \mid r_{i,j} \leq A\sqrt{I} \text{ for } t \in [t_{i,1}, t_{i,2}]\}$: this partition depends on the choice of t_i. To determine the behavior of the center of mass, $\mathbf{c}_s$, of each group where M_s is the total mass of the group, notice that

$$M_s \mathbf{c}_s'' = \sum_{j,k \in G_s} \frac{m_j m_k (\mathbf{r}_j - \mathbf{r}_k)}{r_{j,k}^3} + \sum_{j=1}^{p} \sum_{i \in G_j, k \in G_s} \frac{m_j m_k (\mathbf{r}_i - \mathbf{r}_k)}{r_{i,k}^3}.$$

The already standard argument shows that the anti-symmetry of the vectors forces the first summation to equal zero while in the second summation each $r_{i,k} \geq 2A\sqrt{I}$. Thus, at least in the interval $[t_{i,1}, t_{i,2}]$, the acceleration $\mathbf{c}_s''$ must be "small;" that is

$$|\mathbf{c}_s''| \leq \frac{B}{I}. \tag{4.76}$$

Intuition

The intuition is that when some particle distances itself from previously close particles, some $\mathbf{c}_s$ must change by a multiple of $\sqrt{I}$. For this to happen with the Eq. 4.76 restricted acceleration, $\mathbf{c}_s'$ must be "large." By using an expression similar to Eq. 4.73, I show that this term cannot be "large enough." That is, by expanding the terms, we have

$$\frac{1}{2}\sum_{s=1}^{p}\sum_{i\in G_s} m_i(\mathbf{r}_i - \mathbf{c}_s)^2 = I - \frac{1}{2}\sum M_s\mathbf{c}_s^2 := I - K. \qquad (4.77)$$

To see why this expression helps find an upper bound on $|\mathbf{c}'_s(t_i)|$, suppose K has a maximum somewhere near t_i. At that time $K'' \leq 0$: by differentiating, this means that

$$\sum M_s {\mathbf{c}'_s}^2 \leq -\sum M_s\mathbf{c}_s \cdot \mathbf{c}''_s. \qquad (4.78)$$

To understand why Eq. 4.78 provides the needed bound on $|\mathbf{c}'_s|$, because each $|\mathbf{c}_s|$ is less than $\sqrt{I}$, the Eq. 4.76 bound shows that at this maximum for K, each $\mathbf{c}'^2$ is bounded above by a multiple of $\frac{1}{\sqrt{I}}$; that is,

$$|\mathbf{c}'_s| \leq \frac{B_1}{I^{1/4}}, \qquad (4.79)$$

where B_1 depends on the masses. Because the initial velocity is small, the acceleration always is small, and the growth in $\mathbf{c}_s$ must be accomplished in a very short time interval, we are describing the impossible. (To avoid the complications of considering the dynamics of a particular group, the argument determines the behavior of all groups.)

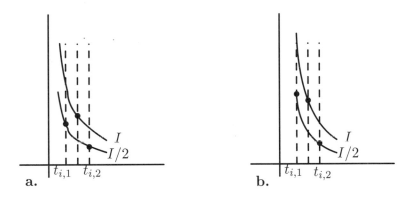

Fig. 4.5. Comparing K with I

A step toward this conclusion is that at t_i the particles in each group are close to one-another, so the left-hand side of Eq. 4.77 has a small value, which means that $K(t_i)$ has nearly the value of $I(t_i)$. In comparison, at times $t_{i,j}$, some terms in this summation on the left-hand side of Eq. 4.77 have large values, which force $K(t_{i,j})$ to be a smaller multiple of $I(t_{i,j})$. In a very short time interval, K grows from a small multiple of I to a larger multiple, and then back to a smaller multiple.

Does this change require K to have a maximum? The problem is illustrated in Fig. 4.5 where two different choice of I and $\frac{1}{2}I$ are plotted. The bullets show where the $K(t)$ curve must pass. In Fig. 4.5a, there is no choice: a smooth K must have a maximum. But with a steeper graph of I, as illustrated in Fig. 4.5b, a curve can be passed through the three points *without* K having a maximum. (We do not need a maximum for K, just that at some point $K'' \leq 0$.) If the I curve is not "too steep," then eith K has a maximum, or it "hesitates."

The idea is to identify the I behaviors that require K to experience a maximum, or at least "hesitate" enough to determine a bound on c'_s. One of several choices is if I is slowly varying. On the other hand, as the sketch in Fig. 4.5b demonstrates, the graph of I must ascend quite rapidly to infinity in order for K to perform all of these rapid changes over a miniscule interval of time without ever experiencing a maximum—this occurs in Xia's construction.

4.9.2   Back to the proof

I now carry out the details of the above outline by first showing that K has a value close to I near $t = t_i$. To do so, notice for each $i \in G_s$, there is a $j \in G_s$ so that $|\mathbf{r}_i - \mathbf{c}_s| \leq |\mathbf{r}_i - \mathbf{r}_j|$. (If G_s is a singleton, $\mathbf{r}_j = \mathbf{c}_s$. To have a singularity, $r_{min} \to 0$, so some G_s has more than one particle.) According to the definition of ϵ, at time t_i, the left-hand side of Eq. 4.77 is bounded above by δI. This means that at t_i, we have that

$$K \geq (1 - \delta)I. \tag{4.80}$$

The next step is to find an estimate for K at the ends of this $[t_{i,1}, t_{i,2}]$ interval: we want to show that K is smaller than a smaller multiple of I. At each endpoint, some mutual distance is $A\sqrt{I}$. Assume that the particles defining this distance are $\mathbf{r}_1$ and $\mathbf{r}_2$. By use of

$$a^2 + b^2 \geq \frac{1}{2}(a - b)^2,$$

we have that the left-hand side of Eq. 4.77 is bounded below by

$$\frac{1}{2}m_0((\mathbf{r}_1 - \mathbf{c}_s)^2 + (\mathbf{c}_s - \mathbf{r}_2)^2) \geq \frac{1}{4}m_0(\mathbf{r}_1 - \mathbf{r}_2)^2 = CI$$

where $C = m_0 A^2/4$ and m_0 is the smallest mass value. That is, at these endpoints, $K \leq (1 - C)I$.

Using these estimates on K at t_i and $t_{i,j}$, $j = 1, 2$, we have that

$$K(t_i) - K(t_i^j) \geq (1 - \delta)I(t_i) - (1 - C)I(t_i^j). \qquad (4.81)$$

The choice of δ (Eq. 4.75) was made to ensure that $(1 - C) < (1 - \delta)$ provides an opportunity for K to have the desired maximum in $[t_{i,1}, t_{i,2}]$. But, as demonstrated with Fig. 4.5, this is true only if I has certain behaviors.

Indeed, with Eq. 4.81, establishing that K has an interior maximum in (t_i^1, t_i^2) if $I \rightarrow L$ is easy. By selecting a sufficiently small value of $t > t_i^2$, we have that $I(t_i)$ and $I(t_{i,j})$ have values sufficiently close to L, which ensure that K has values that are larger than at the endpoints of $[t_{i,1}, t_{i,2}]$. To illustrate with Fig. 4-5a, for sufficiently small values of t, the graph of I is close to horizontal, so K must have a maximum.

In particular, let $\epsilon_1 = \min(L/100, (C - \delta)/100)$ and let t^* be such that for $(0, t^*)$, we have $L - \epsilon_1 < I < L + \epsilon_1$. Selecting t_i so that $t_{i,2} < t^*$ ensures that the right-hand side of Eq. 4.81 is bounded below by

$$(1 - \delta)(L - \epsilon_1) - (1 - C)(L + \epsilon_1) = 0.99(C - \delta)L > 0.$$

This estimate ensures that K has a maximum in the interior of this interval and that some $\mathbf{c}_s$ has experienced a change; i.e., there is a positive value D, depending only on the masses, δ, C, and L, and some s so that

$$|\mathbf{c}_s(t_i) - \mathbf{c}_s(t_{i,2})| > D. \qquad (4.82)$$

While the fact there is a $D > 0$ depending on these variables (but not t_i) suffices for our purposes, the interested reader can use Eq. 4.77 and the definition of K to show that for some s,

$$\mathbf{c}_s^2(t_i) - \mathbf{c}_s^2(t_{i,2}) \geq (C - \delta)\frac{L}{2pM_s} \geq (C - \delta)\frac{L}{2NM} := D\frac{\sqrt{8L}}{m_0}.$$

Because $m_0\mathbf{c}_s^2(t_i) < 2L$, we have Eq. 4.82.

We now have the ingredients: some $\mathbf{c}_s$ changes as in Eq. 4.82 where the change does not depend on the time, the value of any $\mathbf{c}_s''$ is "small" in the $[t_i^1, t_i^2]$ interval (Eq. 4.76), and at some time in this interval $|\mathbf{c}_s'|$ also is small (Eq. 4.79). By integration, we have that

$$|\mathbf{c}_s'| \leq \frac{B_1}{L^{1/4}} + (t_{i,2} - t_{i,1})\frac{B}{L}.$$

This means that for any s, we have

$$|\mathbf{c}_s(t) - \mathbf{c}_s(t_i)| \leq |t - t_i|(\frac{B_1}{L^{1/4}} + (t_{i,2} - t_{i,1})\frac{B}{L}).$$

By selecting a sufficiently small value of t_i, it follows that the inequality of Eq. 4.82 cannot be satisfied, and this completes the proof for $I = O(1)$.

4.9.3 The last steps

It remains to handle the case where $I \to \infty$; this is where, in addition to the possibility of a maximum for K (which would complete the proof), we also have to consider the possibility of K "hesitating" somewhere near t_i. To handle the hesitations, the proof depends on a preliminary result about the behavior of slowly varying functions.

Lemma 4.4 *If I is slowly varying as $t \to 0$, then $tI'/I \to 0$ as $t \to 0$. For $\epsilon > 0$, $0 < \beta < 1$, after some time there is a $\xi \in (\beta t, t)$ so that*

$$|I''(\xi)| < \epsilon I(t)/t^2. \tag{4.83}$$

Equation 4.83 shows that there are times when I'', or from the Lagrange-Jacobi equation, T, cannot be much larger than the value of I. This slowing down of T, or the maximum values of the velocities, is what replaces a maximum point of K.

Proof: Because I is slowly varying, $I(\beta t)/I(t) \to 1$ as $t \to 0$. Using the Taylor series expansion of I about t and $\beta \in (0,1)$, it follows that

$$\frac{I(\beta t)}{I(t)} - 1 = (\beta - 1)\frac{tI'(\zeta)}{I(t)}$$

where $\beta t < \xi < t$. As I is slowly varying, we have that $(\beta - 1)\frac{tI'(\xi)}{I(t)} \to 0$ as $t \to 0$. But we know from the Lagrange-Jacobi equation and the assumption there is a singularity that eventually $|I'|$ is monotonically increasing. (Remember, as the singularity is being approached from $t \to 0+$, $I' < 0$: this is illustrated in Fig. 4.5.) Thus $|I'(\xi)| \geq |I'(t)|$. From this, we have the desired $tI'(t)/I(t) \to 0$ as $t \to 0$.

To prove the second statement, use the Taylor series

$$\frac{I(\beta t)}{I(t)} - 1 = (\beta - 1)\frac{tI'(t)}{I(t)} + (\beta - 1)^2\frac{t^2 I''(\xi)}{I(t)}.$$

In this expression, the left-hand side and the first term on the right-hand side approach zero. Thus, $(\beta - 1)^2\frac{t^2 I''(\xi)}{I(t)} \to 0$, and eventually, in magnitude, this term is less than $(\beta - 1)^2\epsilon$. Collecting terms, this requires

$$|I''(\xi)| \leq \epsilon I(t)/t^2,$$

which is what we wanted to prove. $\square$

As we will see, rather than using the fact I is slowly varying, we use the properties specified in Lemma 4.4. Consequently, the assertion extends to a larger class of functions, such as where $I \sim \frac{L(t)}{t^\alpha}$ where $L(t)$ is slowly varying. However bounds must be imposed on α for the proof to work.

If K'' would ever become non-positive in $[t_i^1, t_i^2]$, then the above proof would apply, and we would be done. Indeed, the proof would be easier because with $I \to \infty$, the bounds on $\mathbf{c}_s'$, $\mathbf{c}_s''$ would be more severe, while some $|\mathbf{c}_s(t_i) - \mathbf{c}_s(t_{i,2})|$ would experience a larger growth: it would be in terms of $D\sqrt{I}$ for some $D > 0$. Indeed, using the same argument that follows Eq. 4.82, we have that there is some s so that

$$|\mathbf{c}_s(t_i) - \mathbf{c}_s(t_{i,2})| \geq (C - \delta)\frac{\sqrt{m_0 I(t_i)}}{8NM}. \tag{4.84}$$

Assume the worse case scenario that K'' remains positive in $[t_{i,1}, t_{i,2}]$: we need to use other approaches to find bounds on $\mathbf{c}_s'$. Here, the continuity of the $r_{j,k}(t)$ distances ensures that there is a small closed interval of time containing t_i so that at each point in this interval, the left-hand side of Eq. 4.77 is bounded above by δI. If the left-hand side of Eq. 4.77 has a minimum in the interior of this interval, call it $t_{i,3}$. If not, then there must be some point in the interior of this interval where the derivative of the left-hand side of Eq. 4.77 is negative. (Remember, I is increasing and that we are using $t \to 0+$, so a negative derivative means that the value is increasing for smaller values of t.)

This value of $t_{i,3}$ is not unique, but whatever the choice, we have that at this point, the left hand side of Eq. 4.77 is bounded above by δI. This allows us to replace t_i in the above analysis, and obtain all of the inequalities, with $t_{i,3}$. The value of this choice of $t_{i,3}$ comes from an additional property: according to Eq. 4.77 and the definition of $t_{i,3}$, we have that

$$|K'(t_{i,3})| \leq |I'(t_{i,3})|. \tag{4.85}$$

The next step is to show there is a distance between $t_{i,3} < t_{i,2}$: this is needed to invoke Eq. 4.83. To explain, it will turn out that the point ξ from Eq. 4.83 will lead to the desired conclusion, but only if ξ is in $(t_{i,1}, t_{i,2})$. To prove this, I establish that for any $\beta \in (0, 1)$, we eventually have that $(\beta t_{i,2}, t_{i,2}) \subset (t_{i,1}, t_{i,2})$. Even stronger, the proof shows that by I being slowly varying, it takes time to even reach $t_{i,3}$: more than $\beta t_{i,2}$. In particular, the claim is that

for any positive $\beta < 1$, and for all t_i sufficiently small, it must be that $t_{i,3} < \beta t_{i,2}$.

If this claim were false, then, for any β, arbitrarily small values of t_i could be found so that $\beta t_{i,2} \leq t_{i,3} < t_{i,2}$. But using $K(t_{i,j}) \leq (1 - C)I(t_{i,j})$ and the Taylor series expansion of K, we have that

$$(1 - C)I(t_{i,2}) \geq K(t_{i,2}) = K(t_{i,3}) + (t_{i,2} - t_{i,3})K'(\xi)$$

where $\xi \in (t_{i,3}, t_{i,2})$.

Remember, $K'' > 0$, so $0 > K'(\xi) > K'(t_{i,3})$. Combining this with $K(t_{i,3}) \geq (1 - \delta)I(t_{i,3})$, Eq. 4.85, the fact $I' < 0$, and the assumption $\beta t_{i,2} \leq t_{i,3}$, we obtain

$$\begin{aligned}(1 - C)I(t_{i,2}) &\geq (1 - \delta)I(t_{i,3}) + (t_{i,2} - t_{i,3})I'(t_{i,3}) \\ &\geq (1 - \delta)I(t_{i,3}) + (\tfrac{1}{\beta} - 1)t_{i,3}I'(t_{i,3}).\end{aligned}$$

By dividing the end expressions by $I(t_{i,3})$, we have

$$(1 - C)\frac{I(t_{i,2})}{I(t_{i,3})} \geq (1 - \delta) + (\frac{1}{\beta} - 1)t_{i,3}\frac{I'(t_{i,3})}{I(t_{i,3})}. \tag{4.86}$$

Here we use the first estimate from Lemma 4.4. By I being slowly varying and the assumptions on $t_{i,2}$, $t_{i,3}$, the left-hand side of Eq. 4.86 approaches $(1 - C)$. For the right-hand side, Lemma 4.4 requires the second term to approach zero. This leads to the contradiction that $(1 - C) \geq (1 - \delta)$, which proves the claim. Thus, for any $\beta \in (0, 1)$, eventually $t_{i,3} < \beta t_{i,2}$.

The rest of the proof uses the kinetic energy T rather than K.[17] To do so, notice that

$$0 \leq \frac{1}{4}\sum_{s=1}^{p} M_s^{-1} \sum_{j,k \in G_s} m_j m_k (\mathbf{r}'_j - \mathbf{r}'_k)^2 = T - \frac{1}{2}\sum_{s=1}^{p} M_s \mathbf{c}'^2_s.$$

Select a positive $\epsilon < \sqrt{m_0}\frac{(C - \delta)}{16NM}$. According to Lemma 4.4, eventually there is a t_i so that a $\xi \in (t_i, t_{i,2})$ can be found that satisfies the inequality $I''(\xi) \leq M\epsilon^2 I(t_{i,2})/4t_{i,2}^2$. According to the Lagrange-Jacobi equation $I'' = T + h$, the fact that $I'' \to \infty$, and the above expression for T, eventually there is a t_i so that, at the $t = \xi$ point established above, we have that

$$\frac{1}{2}\sum M_s \mathbf{c}'^2_s(\xi) \leq T(\xi) = I''(\xi) - h \leq 2I''(\xi) \leq M\epsilon^2\frac{I(t_{i,2})}{2t_{i,2}^2}.$$

---

[17]This choice is for convenience and to indicate other available tools for this kind of analysis: the reader can fashion a proof using K.

The purpose is to find a value for all $\mathbf{c}_s$ at some ξ in the interval; it is $|\mathbf{c}'_s(\xi)| \leq \epsilon \frac{\sqrt{I(t_{i,2})}}{t_{i,2}}$. Using this value with the bound on $\mathbf{c}''_s$, we have that $|\mathbf{c}'_s| \leq \epsilon \frac{\sqrt{I(t_{i,2})}}{t_{i,2}} + B \frac{t_{i,2}}{I(t_{i,2})}$ for $t \in [t_{i,1}, t_{i,2}]$ where the inequality $(t_{i,2} - t_{i,1}) < t_{i,2}$ is used. It now follows that

$$|\mathbf{c}_s(t_i) - \mathbf{c}_s(t_{i,2})| = |(t_{i,2} - t_i)\mathbf{c}'_s(\xi^*)| \leq \epsilon \frac{t_i}{t_{i,2}} \sqrt{I(t_{i,2})} + B \frac{t_i t_{i,2}}{I(t_{i,2})}$$

where $\xi^* \in (t_{i,1}, t_{i,2})$.

By comparing this inequality bounding $|\mathbf{c}_s(t_i) - \mathbf{c}_s(t_{i,2})|$ with the inequality in Eq. 4.84, which provides a lower bound for $|\mathbf{c}_s(t_i) - \mathbf{c}_s(t_{i,2})|$, because $I(t_i) > I(t_{i,2})$ it is clear that once $t_{i,2}$ is sufficiently small a contradiction is obtained. After all, the last term on the right-hand side of the above inequality approaches zero, and the first term is bounded above by $\epsilon \sqrt{I(t_{i,2})}$. Our choice of ϵ was selected to ensure the contradiction eventually occurs. Because $t_i \rightarrow 0$, this must happen. This finally completes the proof of the theorem. $\square$

Chapter 5

How likely is it?

Collisions happen: how likely are they? To the best of my knowledge, J. E. Littlewood was the first to raise this issue. The earliest mention of this problem that I could find is in a footnote in an early edition of Littlewood's delightful *A Mathematician's Miscellany.* Later Littlewood spelled out the issue as Problem #13 in his book *Some Problems in Real and Complex Analysis* [34]. In this latter book, Littlewood explicitly asked whether the set of initial conditions leading to collisions are of Lebesgue measure zero.

Imagine the beautiful paradox that would arise if this were not true. After all, it follows from Sundman's work (Chapter 4) that the set of initial conditions leading to binary collisions form a smooth, lower dimensional submanifold. (The conclusion follows because with Sundman's regularization, binary collisions become regular points.) But a triple or more complicated multiple collision seems to include a binary collision, so imagine the delicious paradox we would have if multiple collisions are more likely! It is almost with disappointment that I report this is not the case. As shown in this chapter, collisions form a set of measure zero and first Baire category.

This likelihood question can be raised with other behaviors. For instance, in Sect. 4.4.3, the possibility of non-collision singularities is discussed. While Xia's clever analysis shows that such motion exists, is it likely? Xia's careful construction leading to a Cantor set of initial conditions appears to constitute rare behavior. What happens in general?

Only a partial answer for this question is known: I will outline my proof showing that motion causing non-collision singularities for the four-body problem is of Lebesgue measure zero. The essense of the argument most surely holds for $N \geq 5$, and I will indicate why, but nobody (to my knowledge) has attempted to carry out the details. While a proof will involve effort, it is reasonable to predict success.

5.1 Motivation

The first of the two main results described in this chapter follows.

Theorem 5.1 *(Saari [77, 78, 83])*[1] *For the coplanar and non-coplanar Newtonian N-body problem, the set of initial conditions leading to a collision, $\mathcal{C}$, forms a set of first Baire category and Lebesgue measure zero. These assertions hold for the lower dimensional setting (of dimension $4(N-1)$ for the coplanar problem and $6(N-1)$ for the non-coplanar problem) where the center of mass is held fixed.*

For the collinear N-body problem, all initial conditions *lead to collisions (either forward or backward in time). However, the set of initial conditions for the collinear problem that leads to a multiple collision where at least one collision involves four or more particles is of Lebesgue measure zero.*

As proved by my student John Urenko [109] in his Ph.D. thesis, the same improbability assertion holds when restricted to a fixed angular momentum variety. Also, Thm. 5.1 can be sharpen to create a distinction among the different kinds of collisions in terms of their Hausdorff dimensions. While the details are left for interested readers, let me recommend using the sharper asymptotic estimates for collisions that involve central configurations. I indicate how this is done in Sect. 5.3.2.

---

[1]My thanks expressed to Carl Simon in [77] reflects an amusing story. Late one night, just before falling asleep, I thought of a way to tackle the Littlewood problem. Sleep was out of the question. Working all night, by morning this approach (which motivated some results in the previous chapter) appeared correct. At this early stage in my career, I had not yet learned to be skeptical of results derived after midnight, so armed with very strong coffee, I drove to campus and excitedly reported my progress to colleagues at Northwestern University. But after some much needed sleep, I discovered a flaw; my result answered the Littlewood problem only for settings with a finite number of central configurations. While this central configuration condition probably is true, it remains an open conjecture. Everyone except Simon, who had left Northwestern, knew of my retraction: in good faith Carl cheerfully publicized my "success" to others. Requests for preprints started rolling in. Driven by acute embarrassment, the kind that is suffered to extremes by young academics, I hurriedly returned to the problem. Fortunately I was able to develop the very different approach described here to finish the problem within two months—in time to respond to preprint requests. So by inadvertently but successfully supplying a powerful drive to quickly resolve the problem, Simon fully deserved my thanks! (Since then Carl only knows about my work that is in final form.)

5.1.1 Idea of proof

The idea of the proof is to start with a sufficiently large set of initial conditions that includes all starting points for collision orbits. The goal is to refine this set by squeezing out those initial conditions whose orbits flirt with colliding, but never make the commitment. This refinement—the squeezing effect—is accomplished by the fact that the N-body problem is a Hamiltonian system, which means that the N-body problem is measure preserving.

To explain what measure preserving means, start with a measurable set of initial conditions $\mathcal{A}_0$. Each initial condition $a \in \mathcal{A}_0$ defines a solution. At some subsequent time t, let $\mathcal{A}_t$ be the set consisting of the location of each of these solutions at time t. By being measure preserving, the measure of $\mathcal{A}_0$ equals that of $\mathcal{A}_t$.

So suppose $\mathcal{C}_{t^*,\mathbf{L}}$ is a set of initial conditions leading to a collision at $\mathbf{L} \in \Delta$ at time $t = t^*$. (The definition of Δ is on page 164.) Just prior to collision, the velocities of the colliding particles approach infinity. On the other hand, the distances from $\mathbf{L}$ to the colliding particles approach zero. This means that as $t \to t^*$, the position and velocity coordinates lie in a rectangle: the length of the rectangle in the velocity direction is very large, but the length in the direction of colliding coordinates is exceedingly small. In particular, the measure of this rectangle $\mathcal{R}(t)$ is bounded by $(t^* - t)^{2/3}(t^* - t)^{-1/3}$ or the vanishingly small value of $(t^* - t)^{1/3}$.

Technical problem require avoiding the continuum of times for the collision. So, the "sqeeze-play" occurs by considering the position of orbits at the different times, $t^* - t = \frac{1}{2^k}$, just prior to collision. Here the collision orbits, and others, are in a box of dimension $(\frac{1}{2^k})^{2/3}(\frac{1}{2^k})^{-1/3}$ just prior to collision. Larger k values, which require the particles to be very close to one another, forces some of the non-colliding orbits out of the box. By using the fact the system is measure preserving and pulling what orbits remain in the box back to the initial conditions, we have that the set of initial conditions with orbits arriving in this box have measure $(\frac{1}{2^k})^{1/3}$.

As $k \to \infty$, orbits flirting with collisions, but not making it, are expelled: only orbits with the serious intent of colliding remain. Thus, the measure of initial conditions that spawn solutions that end up in these skewed rectangular boxes approaches zero. This means that the fixed measure of the initial set of initial conditions leading to collisions must be zero.

While this sketch captures the sense of the proof, complications arise because it is not known *when* or *where* a collisions will occur. Also, the proof uses the constant A from Thm. 4.10, where the value of A is not known. So, details of the proof must describe all possible times of collision, all possible locations for the collisions, all possible "A" multiples for the collision, and the different dimensions of physical space.

5.1.2   Why do we need the Baire category statement?

Because set C has measure zero, it may be surprising to see the accompanying topological assertion about Baire category. Why isn't the Baire category conclusion a direct consequence of the Lebesgue measure result and continuity of solutions with respect to initial conditions? This is almost so, but not quite true.

The category assertion does not follow directly from the probability conclusion because of complications created by possible non-collision singularities. Indeed, the proof about the category conclusion uses bounds on the growth rate of these pathologically appearing singularities—which we now know exist—to show that even in a topological sense, collisions are unlikely.

An interesting challenge is to discover more about the structure of C. To explain, as indicated earlier, there are many cases (based on whether there are a finite number of central configurations) where the initial conditions leading to collisions form smooth lower-dimensional manifolds. Can these sets pile back upon themselves much like the orbits of an irrational flow on a torus? In other words, what happens to the structure of C as $t \to \infty$?

A weaker question is to determine the structure of $\overline{C}$, the closure of C. Does this set have measure zero? An even weaker, but still interesting assertion, would be to show that C is nowhere dense. As described earlier, are the initial conditions for the non-collision singularities in $\overline{C}$?

In other words, a more complete analysis of the structure of C might reveal interesting, and wild, N-body behaviors. The weird and absolutely delightful behavior already admitted by N-body problems suggests that much more is waiting to be discovered: a study of C might uncover some new structures. After all, the behavior of the N-body problem keeps revealing fascinating surprises!

5.2   Proof: C is of first Baire category

The proof (from Saari [83]) uses the facts that set C is of Lebesgue measure zero and that the non-collision singularities force $I \to \infty$ in finite time. This $I \to \infty$ behavior differs from collisions where I approaches a finite limit.

The proof is by contradiction. I show how the assumption that C is not of first Baire category contradicts the fact that C is of measure zero. The first step is to find a subset of C, which also is of second category, with more specified information about when the collisions occur and properties of the collision orbits.

The main tool is the standard "continuity of solutions with respect to initial conditions." This property ensures that if a solution generated by initial condition p exists on a compact time interval, then the solutions for those initial conditions that are sufficiently close to p remain near the original solution.

5.2.1 Finding an appropriate C subset

To start, assume for some N-body problem and choice of masses that C is of second Baire category. First we find a second category subset of C where we know approximately *when* the collisions occur and we have a bound on the size of the moment of inertia I. To do this, let C_m, $m = 0, \pm 1, \pm 2, \ldots$, be the set of initial conditions for which

1. the solution suffers its first singularity in the time interval $[m, m+1]$, and

2. this singularity is caused by collisions.

Of course, $C = \cup C_m$.

Combining the assumption that C is of second category with the fact that C is given by a countable union of the C_m sets means that there is some value of m, say $m = M$, where C_M is of second category. Thanks to the time reversibility of the system, we can assume that $M \geq 0$. Rather than C, now concentrate on the initial conditions in C_M where the collisions occur at some time between M and $M+1$.

The next bookkeeping step is to subdivide C_M; this is done in terms of the maximum size of the moment of inertia I. To do this, let C_M^j be the set of initial conditions p such that

1. $p \in C_M$, and

2. if the solution defined by p has its first singularity at $t^* \in [M, M+1]$, then $I(t) \leq j$ for $0 \leq t < t^*$.

To use this step, which provides information about I, I need to show that $C_M = \cup_{j=1}^{\infty} C_M^j$. One direction is easy: the construction of the C_M^j sets takes points from C_M, so $C_M \supset \cup_{j=1}^{\infty} C_M^j$.

To prove equality, notice that if $p \in C_M$, then its solution has its first singularity—a collision—at some time t^* between M and $M+1$. Moreover, because the singularity is a collision, $I(t)$ has a finite limit as $t \to t^*$. It now follows from continuity considerations that I is bounded above on $[0, t^*]$; this upper bound can be given by some positive integer j. As the initial condition p belongs to $\mathcal{C}_M^j$,

$$\mathcal{C}_M = \cup_{j=1}^{\infty} \mathcal{C}_M^j. \tag{5.1}$$

Again, because $\mathcal{C}_M$ is of second category, there is some choice of j, say $j = J$, so that $\mathcal{C}_M^J$ also is of second category.

It is this set $\mathcal{C}_M^J$ of initial conditions, a set that is of second category and where we have bounds on when the collision occurs and the size of I, that we emphasize.

Properties of $\mathcal{C}_M^J$.

Let $\mathcal{C}_M^{J\,o}$ be the interior of $\overline{\mathcal{C}_M^J}$—the closure of $\mathcal{C}_M^J$. Because $\mathcal{C}_M^J$ is of second category, set $\mathcal{C}_M^{J\,o}$ is a non-empty open set. But by being an open set, $\mathcal{C}_M^{J\,o}$ has a positive Lebesgue measure. Because $\mathcal{C}$ is of Lebesgue measure zero, there must be

$$p \in \mathcal{C}_M^{J\,o} \text{ such that } p \notin \mathcal{C}.$$

But, if initial condition p does not lead to collisions for any time in the interval $[M, M + 1]$, then there are only two possible fates for its solution.

1. The solution defined by p has no singularities in the time interval $[0, M + 1]$.

2. The solution terminates in a non-collision singularity at some time $t^* \in (0, M + 1]$.

The strategy for the rest of the proof is to use the continuity with respect to initial conditions to prove that both cases define an open ball of initial conditions, $\mathcal{B}$, about p where $\mathcal{B}$ is disjoint from our target set $\mathcal{C}_M^J$. This will complete our proof because it means that p cannot be in the closure of $\mathcal{C}_M^J$, so p cannot be in the interior of the closure. This would contradict the existence of p.

If p has no singularities

The first case is where the solution is spared any singularities on the time interval $[0, M + 1]$. From the continuity of solutions with respect to initial conditions, for any $\epsilon > 0$, there exists a open ball of initial conditions about

p with the property that as long as the solutions exist in the time interval $[0, M + 1]$, they differ from the one defined by p by no more than ϵ. If all of these solutions exist on the time interval $[0, M + 1]$, we have the desired ball $\mathcal{B}$. Thus, we just have to establish that all of these solutions exist for the full time interval.

The solution defined by p is bounded on the interval $[0, M+1]$: this is true in the position and velocity components. According to continuity of initial conditions, the bounded solution defined by p forces the other solutions to remain bounded as long as they do not suffer a singularity. But if any of them did encounter a singularity, the solution must be unbounded. Namely, if one of these neighboring solutions has a non-collision singularity, then I, or some position vector, becomes unbounded; if a neighboring solution is a collision singularity, T, and hence some velocity, becomes unbounded. Consequently the continuity with respect to initial conditions provides the open neighborhood $\mathcal{B}$ of p needed to reach a contradiction.

If p causes a non-collision singularity

Now suppose that the solution defined by p dies in a non-collision singularity at some time $t^* \in (0, M + 1]$. As developed and discussed earlier, such a singularity is accompanied by $I \to \infty$ as $t \to t^*$. But if I is reaching for infinity, then at some time t_1 prior to t^*, we have that $I(t_1) > 3(J + 100)$.

The solution corresponding to p exists on $[0, t_1]$. So, for any $\epsilon > 0$, there is an open ball of initial conditions $\mathcal{B}$ about p so that the solution for any initial condition in $\mathcal{B}$ never strays from the solution defined by p by more than ϵ on $[0, t_1]$. To force our contradiction, because the solution corresponding to p has $I > 3(J + 100)$, each of these neighboring solutions in $\mathcal{B}$ must eventually have their I value exceed $2J$ at some time on $[0, t_1]$. But this condition means that $\mathcal{B}$ is disjoint from $\mathcal{C}_M^J$, so it completes our proof about the Baire category. $\square$

5.2.2 A comment about the set of singularities

This same "continuity with respect to initial conditions" argument ensures that the set of singularities leading to any kind of singularity at some time is a F_σ set.

Corollary 5.1 *(Saari [83] Let $\mathcal{S}_N$ be the set of initial conditions leading to a singularity at some time in the interval $[-N, N]$. Set $\mathcal{S}_N$ is a closed set. Consequently, the set of initial conditions leading to a singularity at some time is F_σ.*

Proof: Assume there exists $p \in \overline{S_N}$ where $p \notin S_N$. This means that the solution corresponding to p exists on the time interval $[-N, N]$. Again, according to the continuity of solutions with respect to initial conditions, there must be some open ball $\mathcal{B}$ about p where the corresponding solutions exist on $[-N, N]$. This contradicts the existence of point p. $\square$

5.3 Proof: $\mathcal{C}$ is of Lebesgue measure zero

The proof (from Saari [78]) is divided into two parts. The second part, which requires more delicate estimates, describes what happens for multiple binary collisions. The first part, where cruder estimates suffice, handles all remaining kinds of collisions.

5.3.1 A common collision for $k \geq 3$ particles

First suppose that only particles $m_1, \ldots, m_k$, $k \geq 3$, collide at a common limit point $\mathbf{L}$. It will be clear from the argument that multiple collisions, where one is a k-fold collision, also are improbable. In proving the result, the center of mass integrals are incorporated into the equations of motion by using Jacobi coordinates.

These coordinates, introduced in Sect. 2.5 (page 69), are defined as follows: $\boldsymbol{\rho}_1 = \mathbf{r}_2 - \mathbf{r}_1$, and $\boldsymbol{\rho}_2$ is the vector from the center of mass of $\mathbf{r}_1$ and $\mathbf{r}_2$ to $\mathbf{r}_3$. In the same fashion, $\boldsymbol{\rho}_j$ is the vector from the center of mass of $\mathbf{r}_1, \ldots, \mathbf{r}_j$ to $\mathbf{r}_{j+1}$. By using Jacobi coordinates, we need not worry about the location of the common collision point $\mathbf{L}$. Instead, the k-fold collision occurs when $\boldsymbol{\rho}_1 = \cdots = \boldsymbol{\rho}_{k-1} = \mathbf{0}$.

Because the integrals describing the motion of the center of mass are given by $\sum m_j \mathbf{r}_j = \mathbf{C}t + \mathbf{D}$, it is clear that the behavior of system variables, $\mathbf{R}$ and $\mathbf{V}$, completely determine the values of $\mathbf{C}, \mathbf{D}, \boldsymbol{\rho}_j, \boldsymbol{\rho}'_j$, $j = 1, \ldots, n-1$, and vice versa. By using Jacobi coordinates, we are implicitly assuming that $\mathbf{C}$ and $\mathbf{D}$ are fixed, so our reduced phase space has dimension $6(N-1)$.

According to Thm. 4.10 (page 173), as a collision at $t = t^*$ is approached, the rate of approach of any particle to its limit, whether colliding or not, is $|\mathbf{r}_i - \mathbf{L}_i| < A(t^* - t)^{2/3}$ and $|\mathbf{v}_i| < A(t^* - t)^{-1/3}$. So, from the definition of Jacobi coordinates, there is a positive constant D, which depends only on the mass values and A, so that as $t \to t^*$

$$|\boldsymbol{\rho}_j| \leq D|t^* - t|^{2/3}, \quad |\boldsymbol{\rho}'_j| \leq D|t^* - t|^{-1/3}, \quad j = 1, \ldots, k-1. \qquad (5.2)$$

Let $\mathbb{R}^{6(N-1)}$ be phase space for the Jacobi coordinates, $\mathbb{R}_1^{3(k-1)}$ be the subspace with the $\boldsymbol{\rho}_j$ position vectors for the colliding particles, $\mathbb{R}_2^{3(k-1)}$ be the corresponding subspace for the $\boldsymbol{\rho}_j'$ velocity vectors, and $\mathbb{R}^{6(N-k)}$ be the orthogonal subspace that describes what happens to the non-colliding particles. Let $\mathbf{p}$ be an arbitrary point in $\mathbb{R}^{6(N-k)}$, and let $S(\mathbf{p}, 1)$ be a sphere of radius in $\mathbb{R}^{6(N-k)}$. Let B be an arbitrary positive integer.

Starting the proof

To use the Eq. 5.2 estimates, $|t^* - t|$ is replaced with $2^{-\alpha}$ for some positive integer α. Toward this end, and with a specified time $t_1 \neq 0$, let $B^\alpha(t_1)$ be the set of points in $\mathbb{R}^{6(N-1)}$ with the following properties:

1. Treating a point in $B^\alpha(t_1)$ as an initial condition at time $t = t_1$, the solution exists in the time interval $[0, t_1]$ (or $[t_1, 0]$ if $t_1 < 0$).

2. The components in $\mathbb{R}^{6(N-k)}$ are in the ball of unit radius $S(\mathbf{p}, 1)$.

3. The magnitude of the (position) components in $\mathbb{R}_1^{3(k-1)}$ are bounded above by $B(2^{-\alpha})^{2/3}$.

4. The magnitude of the (velocity) components in $\mathbb{R}_2^{3(k-1)}$ are bounded by $(2^{-\alpha})^{-1/3}$.

To see that $B^\alpha(t_1)$ is a measurable set, notice that conditions 2-4 define a regular object (a product of a sphere with two cubes). According to the "continuity of solutions with respect to initial conditions," the first condition also defines a measurable set: if $\mathbf{p}$ defines a solution on $[0, t_1]$, then (see page 212) so does an open set about $\mathbf{p}$. A computation shows that

$$\mu(B^\alpha(t_1)) \leq E(2^{-\alpha})^{k-1}$$

where E is a positive constant independent of α.

Now consider an arbitrary unit interval of time with rational endpoints, say $[1, 2]$, and divide it into $2^{\alpha+5}$ equal parts. The partition points defines the points $t_0 = 1, \ldots, t_{2^{\alpha+5}} = 2$. Let $B^\alpha[1, 2]$ be the set of initial conditions at time $t = 0$ where the orbit is in $\cup B^\alpha(t_i)$.

As the N-body problem is measure preserving, $B^\alpha(t_1)$ can be pulled back to a set at time $t = 0$ that has the same measure. This means that $B^\alpha[1, 2]$ is measurable (it is the finite sum of measurable sets) and

$$\mu(B^\alpha[1, 2]) \leq E(2^{-\alpha})^{k-2} \tag{5.3}$$

where E is another positive constant that, again, is independent of α.

Next define

$$\mathcal{B}[1,2] = \limsup_{\alpha \to \infty} B^\alpha[1,2].$$

Thanks to the properties of measurable sets, we have that $\mathcal{B}[1,2]$ is measurable. Also, for any n, $\mathcal{B}[1,2] \subset \cup_{\alpha=n}^\infty B^\alpha[1,2]$, so

$$\mu(\mathcal{B}[1,2]) \leq \sum_{\alpha=n}^\infty \mu(B^\alpha[1,2]) \leq E\sum_{\alpha=n}^\infty ((\frac{1}{2})^{k-2})^\alpha. \tag{5.4}$$

For $k > 2$ (and we have $k \geq 3$), the right-hand side is the summation $\sum_{\alpha=n}^\infty x^\alpha$ where $|x| < 1$, so this summation is convergent. Thus, by choosing n sufficiently large, the bound on $\mu(\mathcal{B}[1,2])$ becomes arbitrarily small: this means that $\mu(\mathcal{B}[1,2]) = 0$.

Relating collisions to $\mathcal{B}[1,2]$

We need to show that initial conditions leading to collisions are in $\mathcal{B}[1,2]$. To do so, define $\mathcal{C}([\frac{5}{4}, \frac{7}{4}], \mathbf{p}, B)$ to be the set of initial conditions so that:

1. The first singularity of the solution is a k-fold collision among particles $\mathbf{r}_1, \ldots, \mathbf{r}_k$, and the collision occurs at time $t^* \in [\frac{5}{4}, \frac{7}{4}]$.

2. The value of D, as defined in Eq. 5.2, is less than $B/2$.

3. The limit of the non-colliding particles is in $S(\mathbf{p}, \frac{1}{2})$.

To show that $\mathcal{C}([\frac{5}{4}, \frac{7}{4}], \mathbf{p}, B) \subset \mathcal{B}[1,2]$, let $\mathbf{p}^* \in \mathcal{C}([\frac{5}{4}, \frac{7}{4}], \mathbf{p}, B)$ be an initial condition where the collision occurs at $t^* \in [\frac{5}{4}, \frac{7}{4}]$. Because the time interval $[1,2]$ has been divided into intervals of length $2^{-(\alpha+5)}$ for each α, there is a unique t_k^* so that $t^* \in (t_k^*, t_{k+1}^*]$: we have that $|t^* - t_{k+1}^*| \leq 2^{-(\alpha+5)}$.

We now describe the positions of the colliding particles in terms of these partition points. To do so, we have as $t \to t^*$ that for $j = 1, \ldots, k-1$, all $|\boldsymbol{\rho}_j| < \frac{B}{2}(t^* - t)^{2/3}$. Consequently, for all t sufficiently close to t^*, we have that $|\boldsymbol{\rho}_j(t)| < \frac{3B}{4}(t^* - t)^{2/3}$. This means that for each sufficiently large value of α, there are partition points t_i so that $|\boldsymbol{\rho}_j(t)| < \frac{3B}{4}(t^* - t_i)^{2/3}$ and $|t_{k+1}^* - t_i| \leq 2^{-\alpha}$. For any such t_i,

$$|\boldsymbol{\rho}_j(t)| < \frac{3B}{4}(t^* - t_i)^{2/3} \leq \frac{3B}{4}(t_{k+1}^* - t_i)^{2/3} < B(2^{-\alpha})^{2/3}.$$

A more subtle argument is needed to select a partition point for $\boldsymbol{\rho}_j'$. We have for all α with sufficiently large values that $|\boldsymbol{\rho}_j'(t)| \leq \frac{3B}{4}(t^* - t)^{-1/3}$. Let

t_i^* be the unique partition point where $t_k^* - t_i^* = 2^{-\alpha} - 2^{-(\alpha+5)}$. Eventually (i.e., for a sufficiently large value of α) there is a t_i^* so that for each j

$$|\rho_j'(t_i^*)| \le \tfrac{3B}{4}(t^* - t_i^*)^{-1/3} \quad < \tfrac{3B}{4}(t_k^* - t_i^*)^{-1/3} < \tfrac{3B}{4}(2^{-\alpha} - 2^{-(\alpha+5)})^{-1/3}$$
$$\le \tfrac{3B}{4}(2^{-\alpha})^{-1/3}(1 - 2^{-5})^{-1/3} \le B(2^{-\alpha})^{-1/3}.$$

The same argument holds for ρ_j, $j = k, \ldots, N-1$. Because $\mathbf{r}_{j+1}$ does not collide, $\rho_j'' = O(1)$ for $j = k, \ldots, N-1$, and ρ_j', ρ_j approach limits of order $O(t^* - t)$. Thus for each sufficiently large value of α, the orbit associated with $\mathbf{p}^*$ is in some $B^\alpha(t_i^*)$, so the initial condition is in $B^\alpha[1,2]$, which in turn means it is in $B[1,2]$. Thus, $\mathcal{C}([\tfrac{5}{4}, \tfrac{7}{4}], \mathbf{p}, B) \subset \mathcal{B}[1,2]$ In turn, this means that $\mathcal{C}([\tfrac{5}{4}, \tfrac{7}{4}], \mathbf{p}, B) \subset \mathcal{B}[1,2]$ is a set of measure zero.

To complete the proof, take the union of all sets such as $\mathcal{C}([\tfrac{5}{4}, \tfrac{7}{4}], \mathbf{p}, B)$ where $\mathbf{p}$ is replaced with a point with rational components, B varies over all positive integers, and the time of the collision $[\tfrac{5}{4}, \tfrac{7}{4}]$ is replaced by all intervals of length one-half where the endpoint is a rational number. All collisions of this k-fold type must be in this countable union of sets of measure zero. Thus, this set is of measure zero.

5.3.2 Lower dimensions, binary collisions, and other force laws

The key to the proof is Eq. 5.4, which shows that the improbability conclusion follow if $\mu(B[1,2]) \le F((\tfrac{1}{2})^\alpha)^g$ where g is a positive number. Any setting with a similar construction and $g > 0$ leads to the same conclusion.

Indeed, mimicking the above argument with a d-dimensional space and a k-fold collision, we find that $\mu(B[1,2]) \le E(2^{-\alpha})^{\frac{d(k-1)}{3} - 1}$, so $g = \frac{d(k-1)}{3} - 1$. With the coplanar problem where $d = 2$, we have that a k-fold collision for $k \ge 3$ is unlikely. But for the collinear N-body problem, where $d = 1$, we need to jump to $k \ge 5$ before we obtain a positive g. A bit later I show how to reduce this comment to $k \ge 4$.

Looking at this expression in a different manner, if $k = 2$, then $g = \tfrac{d}{3} - 1$, so trouble arises when using this approach for a binary collision for any dimensional space where $d < 4$.[2] Binary collisions are improbable for $d = 2, 3$, we just need to develop a sharper argument. A likely candidate for improved estimates are the velocity terms. As shown next, by using sharper velocity estimates, the proof is completed.

---

[2]We do not need to prove the result for a binary collision because we know the set of initial conditions is in a lower dimensional smooth submanifold. But, the challenge is here, so why not answer it.

Binary collisions

I first show that a binary collision is improbable. The argument replaces the crude $|\mathbf{r}'_i(t)| \leq D(t^* - t)^{-1/3}$ estimate with a sharper one. Toward this end, recall that a binary collision between $\mathbf{r}_1, \mathbf{r}_2$ satisfies

$$\mathbf{r}_i - \mathbf{L}_i = \mathbf{c}_i(t^* - t)^{2/3} + O((t^* - t)^{5/3}), \ \mathbf{v}_i = \frac{2}{3}\mathbf{c}_i(t^* - t)^{-1/3} + O((t^* - t)^{2/3}), i = 1, 2,$$
$$(5.5)$$

where the $\mathbf{c}_i$'s satisfy the central configuration equation

$$-\frac{2}{9}\mathbf{c}_i = \frac{m_j(\mathbf{c}_j - \mathbf{c}_i)}{c_{ij}^3}.$$

(See, for example, Thm. 4.18, page 193.) The following proof uses the cruder estimates $\mathbf{r}_i - \mathbf{L}_i = \mathbf{c}_i(t^* - t)^{2/3} + o(t^* - t)$, $\mathbf{v}_i = \frac{2}{3}\mathbf{c}_i(t^* - t)^{-1/3} + O(1)$.

Up to rotation, the $\mathbf{c}_1, \mathbf{c}_2$ terms are uniquely determined by the masses, where $\mathbf{c}_1 = -\lambda\mathbf{c}_2$ and λ is uniquely determined by the masses. (See Chap. 3.) Thus

$$\boldsymbol{\rho}_1 = \mathbf{d}(t^* - t)^{2/3} + o(t^* - t), \quad \boldsymbol{\rho}'_1 = \frac{2}{3}\mathbf{d}(t^* - t)^{-1/3} + O(1)$$

where $\mathbf{d} = \mathbf{c}_2 - \mathbf{c}_1$ is uniquely determined up to rotation.

Let $\mathcal{S}_1(\alpha)$ be the sphere in the above $\mathbb{R}_1^d$ (for $\boldsymbol{\rho}_1$) of radius $(2^{-\alpha})^{2/3}|\mathbf{d}|$ and $\mathcal{S}_2(\alpha)$ be the sphere in the above $\mathbb{R}_2^d$ (for $\boldsymbol{\rho}'_1$) with radius $\frac{2}{3}|\mathbf{d}|2^{\alpha/3}$. Let the definition of $B^\alpha(t_i)$ be the same as above (page 215) for conditions 1-2. Replace conditions 3 and 4 with

1. the components in $\mathbb{R}_1^d$ are within distance $(2^{-\alpha})$ of $\mathcal{S}_1(\alpha)$.

2. the components in $\mathbb{R}_2^d$ are within distance $(2^{-\alpha})^{-.01}$ of $\mathcal{S}_2(\alpha)$.

To compute $\mu(B^\alpha[1,2])$, notice that the $\boldsymbol{\rho}_1$ component lies near the surface of the sphere $\mathcal{S}_1(\alpha)$. As the area of the sphere is a multiple of $(2^{-2\alpha/3})^{d-1}$, the volume of the region is a multiple of

$$(2^{-2\alpha/3})^{d-1}2^{-\alpha} = (2^{-\alpha})^{\frac{2d+1}{3}}.$$

The $\boldsymbol{\rho}'_1$ term is near the surface of the $\mathcal{S}_2(\alpha)$ sphere. As the area of this sphere is a multiple of $(2^{\alpha/3})^{d-1}$, the volume is a multiple of

$$(2^{\alpha/3})^{d-1}(2^{-\alpha})^{-.01} = (2^{-\alpha})^{\frac{1-d}{3}-0.01}.$$

This means that

$$\mu(B^\alpha(t_i)) \le E(2^{-\alpha})^{\frac{d+2}{3}-0.01}.$$

In turn, we have that

$$\mu(B[1,2]) \le E(\frac{1}{2}^\alpha)^{\frac{d+2}{3}-1.01},$$

which means that the $g = \frac{d+2}{3} - 1.01 > 0$ for $d = 2, 3$. This is what we wanted to prove.

Some remaining issues

Collinear central configurations are well defined, so similar estimates can be found for the k-fold collisions in the collinear N-body problem. By using these estimates and the above approach, it can be shown that such collisions are improbably for $k \ge 4$. It may be that this estimate can be made sharper; e.g., I have not checked to see whether triple collisions in the collinear problem are unlikely.

It now is easy to show that multiple simultaneous collisions are unlikely. I illustrate with two simultaneous binary collisions. The difference is that the location of the second collision must be specified. So, consider cluster variables where ρ_2 is the vector from the center of mass of the first colliding binary to the center of mass of the second; let ρ_3 be the position of m_5 relative to the center of mass of the first four particles, and continue with the Jacobi coordinates. Then, let $\mathbf{r}_3, \mathbf{r}_4$ be the position of particles three and four *relative to* ρ_2. In this setting, we need to specify the limiting position of ρ_2, so let this be a point in a d-dimensional cube $\mathcal{L}$. The volume of the region defined by the second set of colliding particles is determined by one of the particles and the center of mass of the system, so it is a multiple of $(\frac{1}{2}^\alpha)^{\frac{2d}{3}}(\frac{1}{2}^\alpha)^{-d/3} = (\frac{1}{2}^\alpha)^{\frac{d}{3}}$. In other words, the volume is smaller. (With sharper estimates on the collision behavior, this number can be made even smaller.)

Other force laws

What about other force laws? For $q \ge 3$, forget such results: as discussed several times earlier in this book, the set of initial conditions leading to collisions includes a non-empty open set. But for $q < 3$, other results follow.

Following the approach used to prove Thm. 4.10, we have for collisions in the inverse q-force law, $1 < q < 3$ (where $q = 2$ is the standard Newtonian choice), that as $\mathbf{r}_i \to \mathbf{L}_i$, there exists a positive constant D so that eventually

$$|\mathbf{r}_i - \mathbf{L}_i| \leq D|t^* - t|^{2/(q+1)}, \quad |\mathbf{r}_i'| \leq D|t^* - t|^{(1-q)/(q+1)}.$$

With these estimates, a similar argument can be constructed. What is the final result? I believe that the set of initial conditions leading to collisions is of measure zero for $q < 3$, but I have not proved this. (It is true if there are a finite number of central configurations: to prove this, use an approach similar to Chap. 4 to obtain sharper asymptotic estimates.) What I can prove is that:

Theorem 5.2 *(Saari [78]) In the inverse q-force law where $q < 17/7$ and where physical space is three-dimensional, the set of initial conditions leading to collisions has measure zero.*

The proof in [78] has two parts. The first is to derive the behavior of collisions for the $q = -1$ force law. From the resulting asymptotic estimates, the answer follows by using the above kind of analysis. The second part uses the above estimates for $1 < q < 3$.

My belief that collisions are unlikely for force laws right up to the dividing choice of $q = 3$ is supported, in part, by the following:

Theorem 5.3 *(Saari [78]) In the inverse q force law where $q < 3$, the set of initial conditions leading to a binary collision, or a multiple collision that includes at least one binary collision, has measure zero.*

5.4   Likelihood of non-collision singularities

To show that non-collision singularities are improbable, an argument similar to the above is fashioned. Currently, all that is known is the following.

Theorem 5.4 *(Saari [84]) For the two or three dimensional four-body problem, the set of initial conditions leading to a non-collision singularity is of measure zero and first Baire Category*

An immediate corollary is that for the two or three dimensional two, three, and four body problems that singularities of any kind are unlikely. The same conclusion most surely extends to all $N \geq 5$, but the details have never been carried out. I will indicate what needs to be done.

The intuition for four bodies is simple. Repeating some of the arguments of the last section of Chap. 4, we know that each of the four particles must interact with others. To see this, if the distance between a pair of particles approaches zero and the others remain separated even by an ϵ distance, then the center of mass of this pair is bounded away from the other two particles. As this requires the acceleration of this center of mass to be bounded, the center of mass approaches a distinct limit. Thus, all particles approach a distinct limit. (This is described in Saari [80, 84].)

It is not difficult to extend this argument to show that infinitely often particles must approach other particles, infinitely closely (because with an ϵ separation, the above argument holds), and then separate from one another. But if the particles separate, something must be done to the escaping particles to cause them to turn around and return for the next interaction. According to the laws of motion, this "something" must be an interaction with another particle. In other words, infinitely often three particles are close to one another while the fourth particle is a distance away. (Recall, $I \to \infty$, so the maximum spacing must approach infinity.)

So, at any instant of time, two particles must be far apart (because $I \to \infty$). Infinitely often either one particle leaves a triplet to visit the distant particle, or a pair leaves the triplet to make this visit. In either case, as soon as the triplet is disbanded, the acceleration on the commuting particle, or the acceleration on the center of mass of the commuting pair, becomes bounded, which allows only limited changes on the velocity vector. In other words, the particle must be aimed quite accurately at the distant particle: with even a slight error it will pass the distant particle never to return to create a non collision interaction.

It is this needed accuracy in aiming at the distant particle that causes all of the motion to rapidly approach a straight line. Estimates on how fast the particles approach the line versus how fast the line can expand and the behavior of the velocities are used, in a manner similar to the above, to prove the theorem.

The same behavior occurs for $N \geq 5$. The key fact is that, infinitely often, particles *must* leave some cluster of particles to visit other clusters. To do so, they must leave one cluster with a carefully aimed direction toward a target cluster. Because all of the motion quickly is defined by the centers of mass of the clusters, the non-collision motion must approach, quite rapidly, a lower dimensional hyperplane. So, the measure-preserving estimates are made in terms of how fast the hyperplane is approached, and how expansive is the motion in the hyperplane. This is how the result could be proved, but I know of nobody who has tried to carry out the details. At least for $N = 5$, this should be a reasonably straight-forward project.

Bibliography

[1] Albouy, Alain, The symmetric central configurations of four equal masses, *Cont. Math.* **198** (1996), 131-135.

[2] Alligood, K., T. Sauer, J. Yorke, *Chaos: An Introduction to Dynamical Systems*, Springer-Verlag, New York, 1996.

[3] Anosov, D., Smooth dynamical systems, *Amer. Math. Soc. Transl.* **125** (1985), 1-20.

[4] Barrow-Green, June, *Poincaré and the Three Body Problem*, American Mathematical Society, Providence, RI., 1997.

[5] Boas, R. P., A Tauberian theorem connected with the problem of three bodies, *Amer. J. Math.* **61** (1939), 161-164.

[6] Bohr, H., Zur Theorie der fastperiodischen Funktionen, I, II, III *Acta Math* **45** (1924) 29-127, **46** (1925), 101–214, **47** (1926), 237-281.

[7] Bohr, H., *Fastperiodische Funktionen*, Springer, Berlin, 1932; translated by H. Cohn, *Almost periodic functions*, Chelsea Pub. Co., 1947.

[8] Cabral, H., On the integral manifolds of the N-body problem, *Invent. Math* **20** (1973), 59-72.

[9] Chazy, J., Sur les singularité impossible du probleème des n corps, *C. R. Acad. Sci.* Paris, Hebdomadaires des Séances **170** (1920), 575-577.

[10] Chazy, J., Sur l'allure du mouvement dans le probleme des trois corps quand le temps coroit indefinement, *Ann. Sci. École Norm.* **39** (1922), 29-130.

[11] Chenciner, A., R. Cushman, C. Robinson, Z. Xia, *Celestial Mechanics: Dedicated to Donald Saari for his 60th Birthday*, Contemporary Mathematics No. 292, AMS, Providence, 2002.

[12] Delgado, J., F. Diacu, E. Lacomba, A. Mingarelli, V. Mioc, E. Perez, C. Stoica, The global flow of the Manev problem, *J. Math. Physics* **37** (6) (1996), 2748-2761.

[13] Devaney, R., Triple collisions in the planar isosceles three body problem, *Invent. Math.* **60** (1980), 249-267.

[14] Devaney, R., *An Introduction to Chaotic Dynamical Systems,* Perseus Pub. Co., 1989

[15] Diacu, F., Near-collision dynamics for particle systems with quasihomogeneous potentials, *Jour. Diff. Eqs.* **128** (1996), 58-77.

[16] Diacu, F., and P. Holmes, *Celestial Encounters*, Princeton University Press, Princeton, 1996.

[17] Diacu, F., E. Perez-Chavela. and M. Santoprete, Saari's Conjecture of the N-Body Problem in the Collinear Case, *Transactions American Math. Society* (to appear).

[18] Dziobek, O., Ueber einen merkwurdigen fall des vielkorperproblems, *Astr. Nachr.* **152** (1900), 33-46.

[19] Easton, R., *Jour. Diff. Eq.* **10** (1971), 371

[20] Feller, W., *An Introduction to Probability Theory and its Applications.* Vol. 2, New York: J. Wiley 1966.

[21] Gerver, J., The existence of pseudocollisions in the plane, *Jour. Diff. Eqs.* **89** (1991), 1-68.

[22] Goldstein, H., *Classical Mechanics,* Addison-Wesley, Reading, Massachusetts, 1950.

[23] G. Hardy, Littlewood, J. E., and G. Pólya, *Inequalities*, Cambridge University Press, Cambridge, UK, 1934.

[24] Hampton, M., and R. Moeckel, Finiteness of relative equilibria of the four-body problem, University of Minnesota preprint, July, 2004.

[25] Hernández-Garduno, A, J. Lawson, J. E. Marsden, Relative equilibria for the generalized rigid body, *Journal of Geometry and Physics* in press.

[26] Hille, E., *Lectures on Ordinary Differential Equations*, Addision-Wesley, Reading, Mass., 1969.

[27] Hulkower, N., Lecture notes from a course on the N-body problem, Northwestern University, 1970,

[28] Hulkower, N., The zero energy three body problem, *Indiana Univ. Math J.* **27**, (1978), 409-448.

[29] Katok, A., and B. Hasselblatt, *Introduction to the Modern Theory of Dynamical Systems*, Cambridge University Press, New York, 1995

[30] Kustaanheimo, P., *Spinor regularization of the Kepler motion,* Ann. Univ. Turku, Scr. AI. **73** (1964).

[31] Kustaanheimo, P., and E. Stiefel, Perturbation theory of Kepler motion based on spinor regularization, *J. Reine Angew. Math.* **218** (1965), 204-219.

[32] Kyner, T., Passage through resonance, in *Periodic Orbits, Stability, and Resonances*, ed.,G. Giacaglia, D. Reidel Publ. Dordrecht, Holland, 1970.

[33] Li, T-Y, J. Yorke, Period three implies chaos, *American Mathematical Monthly 82* (1975), 985-992.

[34] Littlewood, J., *Some problems in Real and Complex Analysis*, Heath, Boston, MA. 1969.

[35] Lehmann-Fllhés, R., *Astr. Nachr.*, **127** (1891), 137-144.

[36] Levi, M., *Qualitative Analysis of the Periodically Forced Relaxation Oscillations*, Memoirs of the AMS, **32**, AMS, Providence, 1981.

[37] Levi, M., A period-adding phenomenon, *SIAM J. Appl. Math* **50** (1990), 943-955.

[38] Levi-Civita, T., Sur la regularisation du probleme des troi corps, *Acta Math.*, **42** (1920) 99-144.

[39] Llibre, J., and E. Pina, Saari's Conjecture holds for the planar 3-body problem, 2002 preprint.

[40] Marchal, C., and D. G. Saari, On the final evolution of the n-body problem, *J. Diff. Equations* **20** (1976), 150–186.

[41] Marchal, C., and D. G. Saari, Hill regions for the general three body problem, *Celestial Mechanics* **12** (1975), 115–129.

[42] Mather, J. and R. McGehee, Solutions of the collinear four-body problem which become unbounded in finite time, *Lecture Notices in Physics* **38** (1975), 573-597.

[43] McCord, C., Saari's Conjecture for the planar three-body problem with equal masses, 2002, preprint.

[44] McGehee, R., Triple collisions in the collinear three body problem, *Invent. Math.* **27** (1974), 191-227.

[45] McGehee, R., Triple collision in Newtonian gravitational systems, *Dynamical Systems Theory and Applications* (J. Moser, ed.), Berlin: Springer-Verlag, 1975, 550-572.

[46] Mittag-Leffler, G., Zur Biographie von Weierstrass, *Acta* **38** (1912), 29-65.

[47] Moeckel, R., Orbits of the three body problem which pass infinitely close to triple collision, *Amer. J. Math.* **103** (1981), 1323-1341.

[48] Moeckel, R., Linear stability analysis of some symmetrical classes of relative equilibria, *Hamiltonian Dynamical Systems: History, Theory, and Applications*, ed. H. Dumos, et. al, IMA **63** Springer-Verlag, New York, 1995.

[49] Moeckel, R., Generic finiteness for Dziobek configurations, *Transactions of AMS*, **353** (2001), 4673-4686.

[50] Moeckel, R., A computer assisted proof of Saari's Conjecture for the planar three-body problem, to appear in *Transactions of the AMS*

[51] Moulton, F. R., The straight linc solutions of the problem of N-bodics, *Annal of Math.* Second Series, **12** (1910), 1-17.

[52] Moulton, F. R., *An Introduction to Celestial Mechanics*, Dover, New York 1970.

[53] Murdock, J., *Celestial Mechanics* **18** (1978). 361-375.

[54] Newcomb, S., Modern Mathematical Thought, *Bull of New York Math Soc.* **4** (1893), 95-107. (Also see D. Saari, *The Way It Was: Early Mathematics from the Bulletin*, AMS, Providence, 2003.

[55] Palmore, J., Measure of degenerate relative equilibria, I, *Ann. of Math.* **104** (1976), 421-431.

[56] Painlevé, P., *Lecons sur la théorie analytic de equations différentielles,* Hermann, Paris, 1897.

[57] Peterson, I., *Newton's Clock,* Freeman and Co., New York, 1993.

[58] Perez, E., D. G. Saari, G. Susin, J Yan, Central Configurations in the charged three-body problem, in *Hamiltonian Dynamics and Celestial Mechanics,*, ed. D. Saari, Z. Xia, AMS, vol 198, 1996

[59] Pizzetti, P., Case Particolari del problema dei tre corpi, *Redniconti della Reale Accademia dei Lincei,* **13** (1904), 17-26.

[60] Pollard, H., *Celestial Mechanics,* Prentice Hall, Inc., New Jersey, 1966. Reprinted as Carus Math Monograph #18, Mathematical Association of America, 1976.

[61] Pollard, H., "The behavior of gravitational systems" *Jour. Math. Mech.* **17** (1967), 601-612

[62] Pollard, H., "Some non-linear Tauberian theorems," *Proc. Amer. Math. Soc.* **18** (1967), 399-401.

[63] Pollard, H., A sharp form of the Virial Theorem, *Bulletin of the AMS,* LXX (1964), 703-5.

[64] Pollard, H., and D. G. Saari, Singularities of the n-body problem I, *Arch. Rational Mech. Anal.* **30** (1968), 263–269.

[65] Pollard, H., and D. G. Saari, Singularities of the n-body problem II, in *Inequalities* II, ed. O. Shisha, Academic Press (1970), 255-259.

[66] Roberts, G., A continuum of relative equilibria in the five-body problem, *Physica D* **127** (1999), 141-145.

[67] Roberts, G., Some counterexamples to a Generalized Saari's Conjecture, 2003, preprint.

[68] Robinson, C. *Dynamical Systems,* CRC Press, Inc., Boca Raton, FL., 1995.

[69] Robinson, C. *Introduction to Dynamical Systems: Discrete and Continuous* Prentice-Hall, 2004.

[70] Robinson, C., and J. Murdock, Some mathematical aspects of spin-orbit resonance. II, *Celestial Mechanics* **24** (1981), 83-107.

[71] Robinson, C., and D. G. Saari, N-body spatial parabolic orbits asymptotic to collinear central configurations, *Jour. Diff. Eq.* **48** (1983), 434-459.

[72] Saari, D. G., *Singularities of the Newtonian n-body problem*, Ph.D. Dissertation, (advisor: H. Pollard), Purdue University, 1967.

[73] Saari, D. G., Some large 0 nonlinear Tauberians theorems, *Proc. Amer. Math. Soc.* **21** (1969), 459-462.

[74] Saari, D. G., On bounded solutions of the n-body problem, in G. Giacaglia (ed), *Periodic Orbits, Stability and Resonances,* D. Riedel, Dordrecht, 1970, pp 76-81.

[75] Saari, D. G., Expanding gravitational systems, *Trans. Amer. Math. Soc.* **156** (1971), 219-240.

[76] Saari, D. G., On oscillatory motion in the problem of three bodies, *Celestial Mechanics* **1** (1970), 343-346.

[77] Saari, D. G., Improbability of collisions in Newtonian gravitational systems, *Trans. Amer. Math. Soc.* **162** (1971), 267-271.

[78] Saari, D. G., Improbability of collisions in Newtonian gravitational systems II, *Trans. Amer. Math. Soc.* **181** (1973), 351-368.

[79] Saari, D. G., Singularities and collisions of Newtonian gravitational systems, *Arch. Rat. Mech. & Math. Anal.* **49** (1973), 311-320.

[80] Saari, D. G., On oscillatory motion in gravitational systems, *J. Diff. Equations* **14** (1973), 275-292.

[81] Saari, D. G., A Tauberian theorem for absolutely continuous functions and for series, *SIAM J. Math. Anal.* **5** (1974), 649-662.

[82] Saari, D. G., Dynamics and clusters of galaxies, pp 273-284 in *Stability of the Solar System and of Small Stellar Systems*, ed. Y. Kozai, IAU, 1974.

[83] Saari, D. G., Collisions are of first category, *Proc. Amer. Math. Soc.* **47** (1975), 442-445.

[84] Saari, D. G., A global existence theorem for the four body problem of Newtonian mechanics, *Jour. Diff. Eqs.* **26** (1977), 80- 111.

[85] Saari, D. G., On the role and properties of N body central configurations, *Celestial Mechanics* **21** (1980), 9-20.

[86] Saari, D. G., Manifold structure for collisions and for hyperbolic-parabolic orbits in the n-body problem, *J. Diff. Eqs.* **55** (1984), 300–329.

[87] Saari, D. G., From rotations and inclinations to zero configurational velocity surfaces I, a natural rotating coordinate system, *Celestial Mechanics* **33** (1985), 299–318.

[88] Saari, D. G., A visit to the Newtonian n-body via elementary complex variables, *Math.Monthly,* **97**, (Feb 1990), 105-119.

[89] Saari, D. G., A chaotic exploration of aggregation paradoxes, *SIAM Review* **37** (1995), 37-52.

[90] Saari, D. G., *Chaotic Expansion of the Newtonian N-body problem*, book manuscript, 2004.

[91] Saari, D., G., *The Way it Was,* American Mathematical Society, Providence, R. I., 2004.

[92] Saari, D. G., and N. Hulkower, On the manifolds of total collapse orbits and of completely parabolic orbits for the n-body problem, *J. Diff. Eqs.* **41** (1981), 27–43.

[93] Saari, D. G., and J. Urenko, Newton's method, circle maps, and chaotic motion, *Amer. Math. Monthly* **91** (1984), 3-17.

[94] Saari, D. G., and Z. Xia, The existence of oscillatory and super hyperbolic motion in Newtonian Systems, *Journ. of Diff. Eq.* **82** (1989), 342-355.

[95] Saari, D. G., and Z. Xia, Off to infinity in finite time. *Notices of AMS,* **42** (May, 1995), 538-546. (Translated version in the Czech journal "Pokroky matematiky, fyziky a astronomie" ("Advances in Mathematics, Physics and Astronomy").)

[96] Saari, D. G., and Z. Xia, *Hamiltonian Dynamics and Celestial Mechanics*, Contemporary Mathematics, **198**, AMS, Providence, RI, 1996.

[97] Santoprete, M., A counterexample to a Generalized Saari's Conjecture with a continuum of central configurations, 2004, preprint.

[98] Saslaw, W. C., *Gravitational physics of stellar and galactic systems*, Cambridge U. Press, Cambridge, 1985.

[99] Sarkovskii, A. N., Coexistence of cycles of a continuous map of a line into itself, *Ukrain. Math. Z.* **16** (1964), 61-71.

[100] Siegel, C. L., Der Dreierstoss, *Ann. Math.* **42** (1941), 127-168.

[101] Slaminka, E., K. Woerner, Central configurations and a theorem of Palmore, *Celestial Mechanics*

[102] Smale, S., Topology and mechanics, II. The planar n-body problem. *Inventiones Math.* **11**, (1970) 45-64.

[103] Smale, S., Mathematical problems for the next century, in *Mathematics: Frontiers and Perspectives*, ed. V. Arnold, M. Atiyah, P. Lax, and B. Mazur, American Math. Soc. 2000, 271-294.

[104] Sperling, H., On the real singularities of the n-body problem, *J. Reine angew. Math.* **245** (1970), 15-40.

[105] Stiefel, E., and G. Scheifel, *Linear and Regular Celestial Mechanics*, Springer-Verlag, New York, 1971.

[106] Sundman, K. F., Recherches sur le problème des trois corps, *Acta Soc. Sci. Rennicae,* **34** (1907) no. 34.

[107] Sundman, K., Le problème des trois corps, *Acta Soc. Sci. Fenn.* **35** (1909).

[108] Urenko, J., Improbability of collisions in Newtonian gravitational systems of specified angular momentum, NU Ph.D. dissertation (advisor: D. Saari), 1975.

[109] Urenko, J., Improbability of collisions in Newtonian gravitational systems of specified angular momentum, *SIAM Journal on Applied Mathematics* **36** (1979), 123-136.

[110] von Zeipel, H., Sur les singularités du problème des n corps, *Ark. Mat. Astron. Pys.* **4** (1908).

[111] Widder, D. V., *The Laplace Transform*, Princeton University Press, 1946.

[112] Wintner, A., *The Analytic Foundations of Celestial Mechanics*, Princeton University Press, 1941.

[113] Xia, Z. *The existence of non-collision singularities in Newtonian systems*, Ph.D. Dissertation (advisor: D. Saari), Northwestern University, 1988.

[114] Xia, Z., The existence of noncollision singularities in Newtonian systems, *Annals of Mathematics* **135**, 411-468.

[115] Xia, Z., Some of the problems Saari did not solve, in *Celestial Mechanics: Dedicated to Donald Saari for his 60th Birthday*, Contemporary Mathematics No. 292, ed. A. Chenciner, R. Cushman, C. Robinson, Z. Xia, AMS, Providence, 2002.

Index

F_δ, 213
Δ, 164
$\mathbf{U}_{rot}$, 71
$\mathbf{U}_{scal}$, 71
$\mathbf{W}_{config}$, 55
$\mathbf{W}_{rot}$, 53
$\mathbf{W}_{scal}$, 53
∇_s, 52
$\tilde{I}$, 92
$\tilde{U}$, 92
$\mathbb{R}^k_+$, 86
$\mathbf{R}$, 51
$\mathbf{V}$, 51

Alligood, K., 14, 223
Almagest, 11
almost periodic, 13
angular momentum, 8, 33
Anosov, D., 169, 223
aphelion, 7
Arecibo, 7
Aristotle, 11, 35

Baire category, 210
Barrow-Green, J., 137, 223
Birkhoff, G. D., 149, 159
black holes, 152
Boas, R., 176, 223
Bohr, H., 13, 223
Bohr, N., 13

Cabral, H., vi, 51, 223
CAL, 104, 110

Caloris Basin, 6
Cantor set, 18
Carlson, D., 26
Cauchy, A., 85, 162
CBMS, v
center of mass, 33, 53
central configuration averaged length, 104
central configurations
    collinear, 90, 95, 119
    collisions, 37
    definition, 35, 40
    dengerate, 46
    Euler similarity classes, 43
    expansion of galaxies, 38
    N-1 dimensional, 89
    nonhomogeneous potentials, 45
    relative equilibria, 38
    relative equilibrium, 31
Chazy, J., 223
Chenciner, A., 223, 231
Collins, S., 26
configurational measure, 40, 42, 43, 55, 62, 65, 76, 156, 157
conjecture, 49, 65
conservation of energy, 34
constraint
    actual values, 126
    using area, 128
    using volume, 127
constraints

degenerate pentahedron, 103, 106
degenerate tetrahedron, 101, 106
degenerate triangle, 100
number, 103
Coulton, P., vi
cracks, 45
Cushman, R., 223, 231

Delgado, J., 224
Devaney, R., 14, 224
Diacu, F., 43, 50, 68, 137, 224
Dziobek, O., 101, 224

Earth
Lagrange points, 89
Eastern Illinois University, vi
Easton, R., 36, 224
Einstein, A., 11, 43, 139
epicycles, 11
Euler's Theorem, 40
Euler, L., 90
Eureka, 89

F ring, 26
Fang-Yen, C., 9
Feller, W., 168, 224

galaxy expansion, 38
Galileo, 26
Galprin, G., vi
Gerver, J., 171, 224

half-astronomical units, 2
Hamilton, W., 142
Hampton, M., 224
Hardy, G., 84, 173, 224
Hasselblatt, B., 189, 225
Helsinki Observatory, 138
Herman, M., vi
Hernández-Garduno, A., 50, 224

Hilbert space, non-separable, 13
Hill curves, 89
Hille, E., 195
Holmes, P., 137, 224
homographic motion, 39
homothetic solution, 40
Hopf maps, 144
Hulkower, N., iii, vi, 192, 225, 229

Inequalities, 84
invariable plane, 33, 158
iterates; convergence, 16
itinerary, 16

Katok, A., 189, 225
Kepler
third law, 5
second law, 8
Kepler equations, 148
Kovalevsky, S., 147
Kustaanheimo, P., 141, 142, 225
Kyner, T., 9, 225

Lacomba, E., 224
Lagrange multipliers, 103
Lagrange points
Earth, 89
Mars, 89
Venus, 89
Lagrange, P., 89, 162
Lagrange-Jacobi equation, 48, 49, 153, 165
generalized, 154
Law of cosines, 130
Lawson, J., 50, 224
Le Verrier, U., 6
Lehmann-Filhés, R., 89, 225
Lennard-Jones force law, 44
Levi, M., 23, 225
Levi-Civita, T., 139, 225
Li, T.Y., 19, 225

linear stability, 132
Littlewood, J. E., 84, 173, 207, 224, 225
Llibre, J., vi, 50, 225

Marchal, C., 149, 225
Mariner 10, 6
Mars
    perceived motion, 2
    Lagrange points, 89
Marsden, J., 50, 224
mass; negative, 117
masses; negative, 112
Mather, J., 169, 226
maximum principle, 162
Maxwell, J. C., 31
McCord, C., 50, 226
McGehee, R., 169, 171, 226
mean
    arithmetic, 85
    geometric, 85
    weighted, 86
mean motion, 8
measure preserving, 209
Mercury
    advance of perihelion, 6, 43, 153
    orbit, 6–10
Mingarelli, A., 224
minimum spacing; r_{min}, 163
Mioc, V., 224
Mittag-Leffler, G., 147, 226
Moeckel, R., 46, 50, 224, 226
Moulton, F. R., 90, 226
Murdock, J., 9, 226, 228

National Bureau of Standards, 44
Neptune, 6
Newcomb, S., 34, 226
Newton's method, 14

Newton, I., 11

Oberwolfach conference, vi
    1970s, 84
    1964, 142
    1978, 144
osculating plane, 65

Pólya, G., 84, 224
Painlevé, P., 164, 165, 227
Palmore, J., 133
Pandora, 27
parabolic motion, 189
Perez, E., 43, 50, 68, 227
perihelion, 6
Peterson, I., 137, 227
Petit, F., 89
Pina, E., 50, 225
Pioneer 11, 26
Pizzetti, P., 56, 227
polar moment of inertia; I, 40, 179
Pollard, H., 7, 44, 167, 172, 173, 175, 177, 227
Prometheus, 26
Ptolemy, 11

quasi-periodic, 13
quaternions, 142

relative equilibria, 38
relative equilibrium, 39
Ricc-Curbatro, 139
Roberts, G., 50, 227
Robinson, C., 9, 14, 186, 189, 223, 227, 231
Rosier, R., vi
Rouche's theorem, 162
Routh, E., 31
rule of signs, 102, 109, 112

Santoprete, M., 50, 68

Sarkovskii sequence, 19
Sarkovskii, A., 19, 230
Saslaw, W., 230
Sauer, T., 14, 223
Scheifel, G., 230
self-potential, 34
sensitivity of initial conditions, 18
Siegel, C., 172, 192, 230
Simon, C., 208
singularity
    algebraic branch point, 193
    characterization, 163
    collision, 163, 164, 208
    complex valued, 148
    definition, 163
    non-collision, 164, 208
    other force laws, 219
slowly varying, 168
Smale, S., 35, 45, 230
SMD, 92
Somigliana-Pizzetti model, 56
space of mutual distances, 92
spectral stability, 132
Sperling, H., 230
spinors, 142
Stiefel, E., 141, 142, 225, 230
Stoica, C., 224
stratified structure, 100
Sundman inequality, 61, 156, 158
Sundman, K., 140, 172, 177, 178,
    207, 230
Susin, G., 227
system
    configurational velocity, 53
    decomposition of $\mathbf{W}_{config}$, 71
    gradient, 52
    inner product, 51
    postion, 51
    rotational configurational ve-
        locity, 64

rotational velocity, 52, 53
scalar configurational velocity,
    64
scalar velocity, 52, 53
velocity, 51

Tauberian Theorem, 173, 174, 177,
    185
Trojan asteroids, 89

universal set, 16
Uranus, 6
Urenko, J., 13, 16, 208, 229, 230

Van der Pol equations, 23
Van der Pol, B., 23
Venus, 9
    Lagrange points, 89
virial theorem, 47, 175
von Zeipel, H., 166, 168, 230
Voyager, 26
Vulcan, 6

Wagon, S., 9
Weierstrass, K., 147
Weierstrass-Sundman theorem, 147,
    150
    proof, 154
Widder, D., 174, 231
Wintner, A., 45, 89, 162, 172, 176,
    231
Woerner, K., 230
Wolf, M., 89
word, 16

Xia, Z., iii, 49, 169, 171, 201, 207,
    223, 229, 231

Yan, J., 227
Yorke, J., 14, 19, 223, 225

Titles in This Series

104 **Donald G. Saari,** Collisions, rings, and other Newtonian N-body problems, 2005

103 **Iain Raeburn,** Graph algebras, 2005

102 **Ken Ono,** The web of modularity: Arithmetic of the coefficients of modular forms and q series, 2004

101 **Henri Darmon,** Rational points on modular elliptic curves, 2004

100 **Alexander Volberg,** Calderón-Zygmund capacities and operators on nonhomogeneous spaces, 2003

99 **Alain Lascoux,** Symmetric functions and combinatorial operators on polynomials, 2003

98 **Alexander Varchenko,** Special functions, KZ type equations, and representation theory, 2003

97 **Bernd Sturmfels,** Solving systems of polynomial equations, 2002

96 **Niky Kamran,** Selected topics in the geometrical study of differential equations, 2002

95 **Benjamin Weiss,** Single orbit dynamics, 2000

94 **David J. Saltman,** Lectures on division algebras, 1999

93 **Goro Shimura,** Euler products and Eisenstein series, 1997

92 **Fan R. K. Chung,** Spectral graph theory, 1997

91 **J. P. May et al.,** Equivariant homotopy and cohomology theory, dedicated to the memory of Robert J. Piacenza, 1996

90 **John Roe,** Index theory, coarse geometry, and topology of manifolds, 1996

89 **Clifford Henry Taubes,** Metrics, connections and gluing theorems, 1996

88 **Craig Huneke,** Tight closure and its applications, 1996

87 **John Erik Fornæss,** Dynamics in several complex variables, 1996

86 **Sorin Popa,** Classification of subfactors and their endomorphisms, 1995

85 **Michio Jimbo and Tetsuji Miwa,** Algebraic analysis of solvable lattice models, 1994

84 **Hugh L. Montgomery,** Ten lectures on the interface between analytic number theory and harmonic analysis, 1994

83 **Carlos E. Kenig,** Harmonic analysis techniques for second order elliptic boundary value problems, 1994

82 **Susan Montgomery,** Hopf algebras and their actions on rings, 1993

81 **Steven G. Krantz,** Geometric analysis and function spaces, 1993

80 **Vaughan F. R. Jones,** Subfactors and knots, 1991

79 **Michael Frazier, Björn Jawerth, and Guido Weiss,** Littlewood-Paley theory and the study of function spaces, 1991

78 **Edward Formanek,** The polynomial identities and variants of $n \times n$ matrices, 1991

77 **Michael Christ,** Lectures on singular integral operators, 1990

76 **Klaus Schmidt,** Algebraic ideas in ergodic theory, 1990

75 **F. Thomas Farrell and L. Edwin Jones,** Classical aspherical manifolds, 1990

74 **Lawrence C. Evans,** Weak convergence methods for nonlinear partial differential equations, 1990

73 **Walter A. Strauss,** Nonlinear wave equations, 1989

72 **Peter Orlik,** Introduction to arrangements, 1989

71 **Harry Dym,** J contractive matrix functions, reproducing kernel Hilbert spaces and interpolation, 1989

70 **Richard F. Gundy,** Some topics in probability and analysis, 1989

69 **Frank D. Grosshans, Gian-Carlo Rota, and Joel A. Stein,** Invariant theory and superalgebras, 1987

68 **J. William Helton, Joseph A. Ball, Charles R. Johnson, and John N. Palmer,** Operator theory, analytic functions, matrices, and electrical engineering, 1987

TITLES IN THIS SERIES

67 **Harald Upmeier,** Jordan algebras in analysis, operator theory, and quantum mechanics, 1987

66 **G. Andrews,** q-Series: Their development and application in analysis, number theory, combinatorics, physics and computer algebra, 1986

65 **Paul H. Rabinowitz,** Minimax methods in critical point theory with applications to differential equations, 1986

64 **Donald S. Passman,** Group rings, crossed products and Galois theory, 1986

63 **Walter Rudin,** New constructions of functions holomorphic in the unit ball of C^n, 1986

62 **Béla Bollobás,** Extremal graph theory with emphasis on probabilistic methods, 1986

61 **Mogens Flensted-Jensen,** Analysis on non-Riemannian symmetric spaces, 1986

60 **Gilles Pisier,** Factorization of linear operators and geometry of Banach spaces, 1986

59 **Roger Howe and Allen Moy,** Harish-Chandra homomorphisms for p-adic groups, 1985

58 **H. Blaine Lawson, Jr.,** The theory of gauge fields in four dimensions, 1985

57 **Jerry L. Kazdan,** Prescribing the curvature of a Riemannian manifold, 1985

56 **Hari Bercovici, Ciprian Foiaş, and Carl Pearcy,** Dual algebras with applications to invariant subspaces and dilation theory, 1985

55 **William Arveson,** Ten lectures on operator algebras, 1984

54 **William Fulton,** Introduction to intersection theory in algebraic geometry, 1984

53 **Wilhelm Klingenberg,** Closed geodesics on Riemannian manifolds, 1983

52 **Tsit-Yuen Lam,** Orderings, valuations and quadratic forms, 1983

51 **Masamichi Takesaki,** Structure of factors and automorphism groups, 1983

50 **James Eells and Luc Lemaire,** Selected topics in harmonic maps, 1983

49 **John M. Franks,** Homology and dynamical systems, 1982

48 **W. Stephen Wilson,** Brown-Peterson homology: an introduction and sampler, 1982

47 **Jack K. Hale,** Topics in dynamic bifurcation theory, 1981

46 **Edward G. Effros,** Dimensions and C^*-algebras, 1981

45 **Ronald L. Graham,** Rudiments of Ramsey theory, 1981

44 **Phillip A. Griffiths,** An introduction to the theory of special divisors on algebraic curves, 1980

43 **William Jaco,** Lectures on three-manifold topology, 1980

42 **Jean Dieudonné,** Special functions and linear representations of Lie groups, 1980

41 **D. J. Newman,** Approximation with rational functions, 1979

40 **Jean Mawhin,** Topological degree methods in nonlinear boundary value problems, 1979

39 **George Lusztig,** Representations of finite Chevalley groups, 1978

38 **Charles Conley,** Isolated invariant sets and the Morse index, 1978

37 **Masayoshi Nagata,** Polynomial rings and affine spaces, 1978

36 **Carl M. Pearcy,** Some recent developments in operator theory, 1978

35 **R. Bowen,** On Axiom A diffeomorphisms, 1978

34 **L. Auslander,** Lecture notes on nil-theta functions, 1977

33 **G. Glauberman,** Factorizations in local subgroups of finite groups, 1977

32 **W. M. Schmidt,** Small fractional parts of polynomials, 1977

31 **R. R. Coifman and G. Weiss,** Transference methods in analysis, 1977

30 **A. Pełczyński,** Banach spaces of analytic functions and absolutely summing operators, 1977

For a complete list of titles in this series, visit the
AMS Bookstore at **www.ams.org/bookstore/**.